Geometric Etudes in Combinatorial Mathematics

Second Edition

Alexander Soifer

Geometric Etudes in Combinatorial Mathematics

Second Edition

With over 350 Illustrations

Forewords by

Philip L. Engel
Paul Erdős
Branko Grünbaum
Peter D. Johnson, Jr.
and Cecil Rousseau

Alexander Soifer
College of Letters, Arts and Sciences
University of Colorado
1420 Austin Bluffs Parkway
Colorado Springs, CO 80918, USA
asoifer@uccs.edu

ISBN: 978-0-387-75469-7 e-ISBN: 978-0-387-75470-3
DOI: 10.1007/978-0-387-75470-3
Springer New York Dordrecht Heidelberg London

Library of Congress Control Number: 2010928053

Mathematics Subject Classifications (2010): 52-01, 52-02, 52A05, 52A10, 52A15, 52A35, 52A37, 05C99, 00A07, 00A08

Cover design: Mary Burgess

Printed on acid-free paper

Springer is part of Springer Science+Business Media (www.springer.com)

In memory of
ISAAC YAGLOM,
the great expositor
of the world of mathematics.

Frontispiece reproduces the front cover of the original edition. It was designed by my late father Yuri Soifer, who was a great artist. Will Robinson, who produced a documentary about him for the Colorado Springs affiliate of ABC Broadcasting Company, called him "the artist of the heart." For his first American one-man show at the University of Colorado in June–July 1981, Yuri sketched his autobiography:

I was born in 1907 in the little village Strizhevka in the Ukraine. From the age of three, I was taught at the Cheder (elementary school by a synagogue), and since that time I have been painting. At the age of ten, I entered Feinstein's Jewish High School in the city of Vinniza. The art teacher, Abram Markovich Cherkassky, a graduate of the Academy of Fine Arts at St. Petersburg, looked at my book of sketches of praying Jews, and consequently taught me for six years, until his departure for Kiev. Cherkassky was my first and most important teacher. He not only critiqued my work and explained various techniques, but used to sit down in my place and correct mistakes in my work until it was nearly unrecognizable. I couldn't then touch my work and continue—this was unforgettable.

In 1924, when I was 17, my relative, the American biologist, who later won the Nobel Prize in (1952) Selman A. Waksman, offered to take me to the United States to study and become an artist, and to introduce me to Chagall, but my mother did not allow this, and I went to Odessa to study at the Odessa Institute for the Fine Arts, in the studio of Professor Mueller. Upon graduation in 1930, I worked at the Odessa State Jewish Theater, and a year later became the chief set and costume designer. In 1934 I came to Moscow to design plays for Birobidzhan Jewish Theater under the supervision of the great Michoels. I worked for the Jewish news paper Der Emes, *the Moscow Film Studio, Theater of Lenin's Komsomol, a permanent National Agricultural Exhibition. Upon finishing 1941–1945 service in the World War II, I worked for the National Exhibition in Moscow VDNH.*

All my life I have always worked in painting and graphics. Besides portraits and landscapes in oil, watercolor, gouache, and marker (and also acrylic upon the arrival in the USA), I was always inspired (perhaps, obsessed) by the images and ideas of the Russian Civil War, World War II, (biblical stories) and the little Jewish village that I came from.

The rest of my biography is in my works!

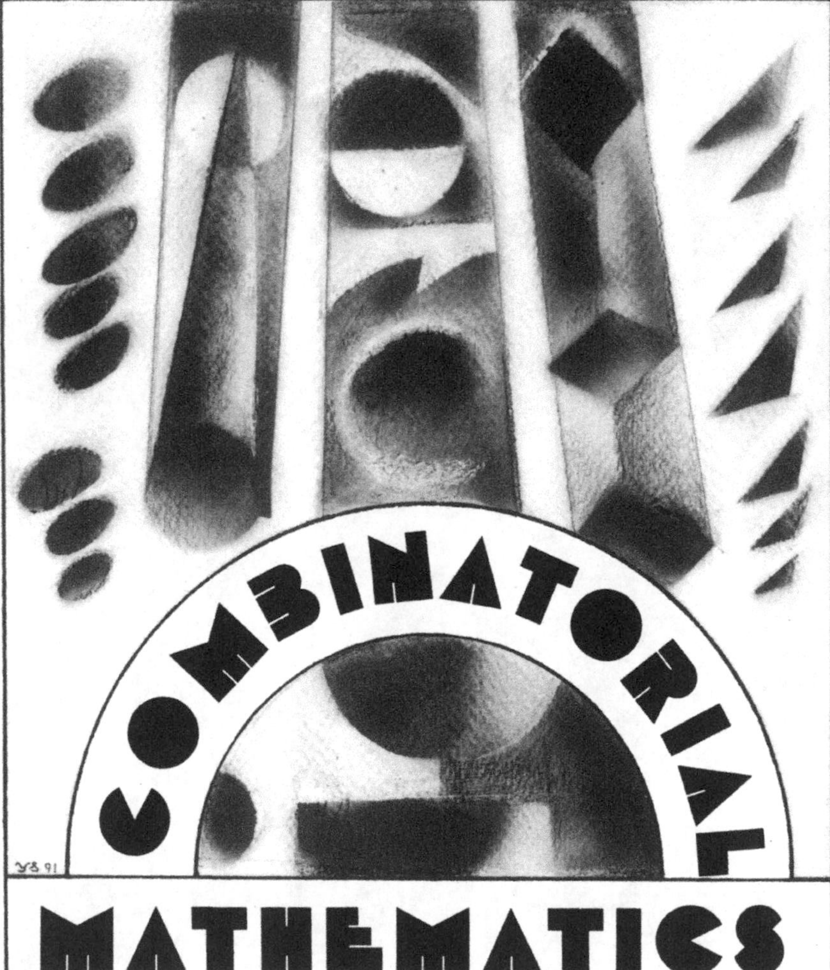

Front cover of the first edition, 1991, by Yuri Soifer.

Forewords to the Second Edition

Each time I looked at *Geometric Etudes in Combinatorial Geometry*. I found something that was new and surprising to me, even after more than fifty years working in combinatorial geometry.

The new edition has been expanded (and updated where needed), by several new delightful chapters. The careful and gradual introduction of topics and results is equally inviting for beginners and for jaded specialists. I hope that the appeal of the book will attract many young mathematicians to the visually attractive problems that keep you guessing how the questions will be answered in the end.

Branko Grünbaum
Professor of Mathematics
University of Washington
September 2008 Seattle, Washington

Vladimir Boltyanski and Alexander Soifer's *Geometric Etudes in Combinatorial Mathematics* was a gem of a book, and it is still there in Soifer's expansion of it, coming out 18 years after the original. What's new are five relatively short but very interesting chapters on developments, over the intervening 18 years, in five areas treated in the original: 2-dimensional tiling, 3-dimensional tiling, Ramsey numbers, the Borsuk Conjecture, and the chromatic number of the plane and its relatives.

You will certainly be interested in any new chapter on a subject that you are already interested in, but you may also find things to interest you in Soifer's writing on subjects with which you are not well acquainted. I am no aficionado of Ramsey numbers, but I found the extra chapter on them very interesting; it was a bit like reading about the discovery of the structure of DNA, or NASA's triumph in the 1960's with the Apollo project. These five chapters are written in Soifer's recently appearing more discursive, anecdotal, historical style, traces of which can be found in the earlier book, but it flows freely here. I like it.

But the heart and neuroskeletal system of the new book is the old book, which cannot be praised enough. Since praise is boring, I will praise it indirectly by speaking of a clash of views within mathematics which concerns how mathematics should be presented, taught, and even practiced.

Timothy Gowers delineated this clash in an insightful essay (that I am not able to lay hands on at the moment, so I warn you that I am working from memory here) in which he pointed out two opposing, or at least different, opinions among mathematicians on what mathematics is, or should be. I will refer to these competing idealizations (because I do not remember how Gowers referred to them) as Mathematics as Problem Solving (which, by the way, happens to be the title of another excellent and useful book by Alexander Soifer) and Mathematics as the Discovery of Structure; MAPS and MADS, for short. Do these terms stand for anything real? I think so, but the memes they stand for are vague, psychological, sociological — as Konrad Lorenz said of aggression in animal behavior, there may not be a neat, comprehensive definition, but you'll know it when you see it. If you are working on operator algebras, you are doing MADS; almost all of combinatorics is MAPS. Paul Erdős did MAPS; Alexandre Grothendieck did MADS.

I agree with Gowers that there is no reason for these views of mathematics to be in competition nor for the MADS aristocracy to look down on the MAPS tribe. There have been great achievements and great mathematicians in each mode (and some, like Gowers and Bollobas, in both), and there are great swatches of mathematics that naturally belong to one mode or the other. We must have them both! But there is at least one important thing that MAPS has that MADS does not: it can be exhilarating fun right from the start, for a student being introduced to mathematics by someone like Soifer, who knows how to go about it. If you want to interest young people in mathematics, MAPS is clearly the way to start.

(A friend who taught for a while in Morocco told me that in the secondary schools there, in an educational system which descended from French colonial rule, a real polynomial was defined to be a finitely non-zero function from the non-negative integers into the reals. Addition of these things was ordinary addition of real-valued functions, and multiplication was given by the obvious convolution. Then everything every schoolboy should know about polynomials was proven, from these definitions. If you think that this is a pretty cool way to do the theory of polynomials — because this way that theory sits naturally within the theory of formal power series, which sits naturally within the theory of formal Laurent series — then you are probably a MADS devotee. If you are, further, thinking of trying out this approach in a course for bright high school students — please, I beg you, don't do it!)

And that brings us back to *Geometric Etudes*. All of Alexander Soifer's books can be viewed as excellent and artful entrees to mathematics in the MAPS mode. Different people will have different preferences among them, but here is something that *Geometric Etudes* does better than the others: after bringing the reader into a topic by posing interesting problems, starting from a completely elementary level, it then goes deep. The depth achieved is most spectacular in Chapter 4, on combinatorial geometry, which could be used as part or all of a graduate course on the subject, but it is also pretty impressive in Chapter 3, on graph theory, and in Chapter 2, where the infinite pigeonhole principle (infinitely many pigeons, finitely many holes) is used to prove theorems in an important subset of the set of fundamental theorems of analysis.

That's enough praising for now. It's a very good book. I hope it finds its way into the hands of youngsters for whom it is primarily intended, and into the hands of their teachers.

Peter D. Johnson Jr.
Professor of Mathematics
Auburn University
October 2008 Auburn University, Alabama

It is generally agreed that the responsibility of a mathematician is to discover and rigorously establish pattern and structure, and that a prerequisite for mathematical permanence is beauty. As G. H. Hardy put it, "The mathematician's patterns, like the painter's or the poet's, must be *beautiful* ... There is no permanent place in the world for ugly mathematics."

Alexander Soifer's *Geometrical Etudes in Combinatorial Mathematics* is concerned with beautiful mathematics, and it will likely occupy a special and permanent place in the mathematical literature, challenging and inspiring both novice and expert readers with surprising and exquisite problems and theorems.

Soifer provides a comprehensive and expertly written introduction to the mathematics of tilings, graphs, colorings, and convex figures, and introduces the reader to the major questions and their framers: Borsuk, Hadwiger, Helly, Jordan, Ramsey, Reuleaux, and Szökefalvi-Nagy. He conveys the joy of discovery as well as anyone, and he has chosen a topic that will stand the test of time.

Cecil Rousseau
Professor of Mathematics
Memphis State University;
Chair, United States of America
Mathematical Olympiad Committee
October 2008 Memphis, Tennessee

Forewords to the First Edition

This interesting and delightful book by two well-known geometers is written both for mature mathematicians interested in somewhat unconventional geometric problems and especially for talented young students who are interested in working on unsolved problems which can be easily understood by beginners and whose solutions perhaps will not require a great deal of knowledge but may require a great deal of ingenuity.

Many unsolved problems are discussed, for example, the tiling of squares with polyominoes, and also many exercises are given of various degrees of difficulty.

There is also an interesting chapter on existence proofs, the understanding of which perhaps requires more mathematical maturity. There is also a chapter on graph theory and a slightly more difficult chapter on the Jordan (Curve) Theorem.

There is also a more difficult chapter on combinatorial geometry where the famous unsolved conjecture of Borsuk is discussed in great detail. Fifty years ago I spent lots of time trying to prove it. To quote Hardy, I hope younger and stronger hands (or rather brains) will have more success.

The last two chapters deal with illumination problems and Helly and Szökefalvi-Nagy's theorem. Here also, many unsolved problems are stated.

I recommend this book very warmly.

Paul Erdős
Member of the Hungarian Academy of Sciences
Honorary Member of the National Academy of Sciences of the USA
January 1991, Gainesville, Florida

How do young people develop skills of any kind — from driving cars to playing basketball or a musical instrument? In all cases the sequence of events is the same: a little instruction, more or less formal, is followed by ample practice. The person wishing to acquire better skills must invest, on his or her own, considerable efforts aimed at gaining better mastery of various aspects of the activity.

Mathematics in general, and geometry in particular, are also fields in which the small amount of formal instruction (often very superficial) given in schools is not sufficient to bring latent talents to full development. Individual work and effort are necessary, and one of the shortcomings of our educational system is the lack of extracurricular material that would make such independent study of geometry (or other mathematics) attractive and interesting.

The present book is an appealing step in the direction of providing useful supplementary reading and practice material. It is intended for the use of our high school students—it was written with that audience in mind, and is aimed at giving accessible but not trivial opportunities for exercising the geometric intuition as well as deductive reasoning. The discussion is self-contained and easy to follow—but despite the elementary character of the mathematics involved, there are plenty of challenging questions, and even several open problems that are easy to state but have so far resisted all attempts to solve them.

The authors build on the tradition and experience of the educational system of the U.S.S.R., in which books of this nature have long played an important role. This represents popularization of science at its best. Many contemporary mathematicians (the writer of these lines included, and still appreciative of the experience) obtained their first taste of the geometry of convex figures from a book very similar in spirit to the present one, which was coauthored by Professor Boltyanski, one of the authors of this work (it is listed under [YB] in the Bibliography).

Throughout the text, the authors show great mastery of the topics discussed. Their infectious enthusiasm for opening our eyes to the beauties of the worlds of geometry and combinatorics should make this book attractive to wide audience. It is to be hoped that the following pages will bring the joy of understanding, seeing and discovering geometry to many of our young people. It is also to be hoped that many other texts of a similar nature will follow, to help lift our teaching out of the present doldrums.

Branko Grünbaum
Professor of Mathematics
University of Washington
February 1991, Seattle, Washington

Some areas of mathematics are not as well known as they deserve to be. They are like the out-of-the-way places where a discerning traveler can find unexpected pleasure and satisfaction. Combinatorial geometry is such a mathematical area. Its basic ideas are easily within the grasp of a bright high school student. However, there are many trained mathematicians who are unaware of even its most essential problems and achievements.

GEOMETRIC ETUDES IN COMBINATORIAL MATHEMATICS provides the reader an opportunity to explore this beautiful area of mathematics. Expertly guided by Alexander Soifer and Vladimir Boltyanski, the reader is surprised and delighted by exquisite gems of geometry and combinatorics. A leisurely and captivating presentation leads the reader into a world of tilings, graphs, and convex figures. It is a world that will be long remembered for its striking problems and results.

Cecil Rousseau
Professor of Mathematics
Memphis State University
Coach of American Team
for the International
Mathematics Olympiad
January 1991, Memphis, Tennessee

Contents

Preface to the Second Edition

> *A mathematician, like a painter or a poet, is a maker of patterns. If his patterns are more permanent than theirs, it is because they are made with ideas. A painter makes patterns with shapes and colours, a poet with words. A painter may embody an 'idea,' but the idea is usually commonplace and unimportant. In poetry, ideas count for a great deal more; but as Housman insisted, the importance of ideas in poetry is habitually exaggerated... A mathematician, on the other hand, has no material to work with but ideas, and so his patterns are likely to last longer, since ideas wear less with time than words.*
>
> *The mathematician's patterns, like the painter's or the poet's, must be* ***beautiful****; the ideas, like the colors or the words, must fit together in a harmonious way. Beauty is the first test: there is no permanent place in the world for ugly mathematics.*
>
> —*G. H. Hardy,* A Mathematician's Apology, *1940* [Har, pp. 24–25]

I grew up on books by Isaac M. Yaglom and Vladimir Boltyanski. I read their books as a middle and high school student in Moscow. During my college years, I got to know Isaak Moiseevich Yaglom personally and treasured his passion for and expertise in geometry and fine art. In the midst of my college years, a group of Moscow mathematicians, including Isaak Yaglom, signed a letter protesting the psychiatric imprisonment of the famous dissident Alexander Esenin-Volpin. Yaglom was fired from his job as professor for that.

In 1970, I visited Yaglom in his downtown Moscow apartment. We discussed problems I had then created about cutting triangles into triangles, which 20 years later became a foundation of my book *How Does One Cut a Triangle?* [S2]. This was an unforgettable mathematical meeting; Yaglom also showed me a powerful oil painting by the Russian avant-garde painter Robert Falk that he owned.

In 1974, the organizers of the *Conference on Mathematical Work with Gifted Students* at Leningrad University scheduled my plenary talk on problems of combinatorial geometry between the talks by Boltyanski and Yaglom. I was humbled to speak between two of the leaders of this field, but in his talk, Yaglom praised my applications of algebraic methods in geometry (on cutting triangles, see [S2] and its expanded edition [S10]); he called them a product of our time that could not have occurred earlier.

I left Russia for the United States in 1978. Shortly after, Yaglom visited my parents. My mother recalled asking him, "Why would you not leave Russia?" "I am too old, and all my friends are here," was Yaglom's answer.

Ten years later, at the 1988 International Congress on Mathematical Education in Budapest, I ran into Vladimir G. Boltyanski who informed me of Yaglom's recent passing on. I asked Boltyanski whether he would like to write a book together and dedicate it to Isaac Yaglom. Boltyanski answered my question with a question: "What do you need me for?" but he added, "Although, it may be more fun to write a book together."

In June of 1990, Vladimir came to Colorado Springs and spent three weeks in my home. As the result of this feverish joint effort, and eight more months on my own, editing and illustrating, the first edition of this book was born. It covered only four chapters out of some twenty-four that we had listed in Budapest as topics of mutual interest, but it was better than nothing, and the first edition appeared in early 1991.

I saw Volodya Boltyanski for the last time in 1993, seventeen years ago in Moscow. His last e-mail arrived from Mexico thirteen years ago, in May of 1997: he lived and worked there, and wanted to come to Colorado Springs to join me to write another book of *Etudes*. We tried, but his notice was too short, and we were unable to arrange Volodya's visit then. This was the last time I heard from him. When in 2007 Springer offered to publish a new expanded edition of this book, I tried to invite Boltyanski to join me in writing it. Regretfully, I did not know his whereabouts. Thus, the job of correcting, updating and substantially expanding this book fell upon me alone. I hope that now, at the age of 85, Volodya is alive and well, and continues to enjoy a healthy and productive life.

In his "extended review" in *The American Mathematical Monthly*, Don Chakerian complimented our choice of *Etudes*:

> Boltyanski and Soifer have titled their monograph aptly, inviting talented students to develop their technique and understanding by grappling with a challenging array of elegant combinatorial problems having a distinct geometric tone. The etudes presented here are not simply those of Czerny, but are better compared to the etudes of Chopin, not only technically demanding and addressed to a variety of specific skills, but at the same time possessing an exceptional beauty that characterizes the best of art . . . Keep this book at hand as you plan your next problem solving seminar.

The great expositor and promoter of this kind of mathematics, Martin Gardner gave *The Etudes* a nod, too:

> Alexander Soifer and Vladimir Boltyanski have produced a fascinating book, filled with material I have not seen before in any book.

For this expanded *Springer* edition, to the original four Chapters, I am adding five new shorter chapters. Let us take a look at their content.

In the eighteen years that followed, one of the many open problems in the book (Problem 5.3) has been solved in 2006 and published in *Geombinatorics* [Ka] by Mitya Karabash, a brilliant undergraduate mathematician from Columbia University (who entered a Ph.D. program of the Courant Institute of the Mathematical Sciences in the fall of 2008). To my surprise, he proved that an $m \times n$ rectangle can be tiled by L-tetrominoes of the same orientation if and only if mn is divisible by 8 and $m, n \neq 1, 3$. Mitya also proved that an $m \times n$ rectangle can be tiled by L-tetrominoes of the same orientation so that the tiling has 2-fold symmetry if and only if mn is divisible by 8 and m, n are both even, or mn is divisible by 16 and $m, n \neq 1, 3$. The new Chapter 5 is dedicated to Mitya's work.

Norton Starr of Amherst College was inspired by Problem 6.10, dealing with packing a parallelepiped with 3-dimensional trominoes, to look into more sophisticated packing of a cube with 3-dimensional trominoes and one 3-dimensional monomino, and determining where the monomino can be placed. Chapter 6 is dedicated to Starr's results, to be published in the October 2008 issue of *Geombinatorics*.

There has been a great progress in determining small Ramsey numbers, much of which was due to works by Geoffrey Exoo, Stanisław Radziszowski and Brendan McKay. Chapter 7 is dedicated to stating some of these results.

As Boltyanski and I predicted in the first edition of this book, the Borsuk Conjecture was disproved by Jeff Kahn and Gil Kalai in 1993 [KK]. This started a competition for a counterexample of the smallest dimension, which is the subject of Chapter 8.

Finding the chromatic number of the plane is my favorite unsolved problem in all of mathematics. Much (although not all) of my new *Mathematical Coloring Book* (published on November 4, 2008 by Springer [S7]) is dedicated to this problem. My desire to include some of my own and others' results in this book is therefore not surprising. They form Chapter 9, the longest of all the new chapters.

I am most grateful to Branko Grünbaum, Peter D. Johnson, Jr., and Cecil C. Rousseau, the first readers of the new manuscript, for their forewords and suggestions.

Love of my children Mark Samuel Soifer and Isabelle Soulay Soifer has been recharging my creative engines and keeping me sane. Shmusik, Belya, I owe you so much!

I am deeply indebted to Ann Kostant for inviting this new expanded edition of the book into the historic Springer. I have been blessed to work with Springer editor Elizabeth Loew – every conversation with her has brightened my day. I thank Susan Westendorf for her help and understanding in supervising production of this book; and Mary Burgess for designing a wonderful cover.

Alexander Soifer
Colorado Springs
September 22, 2008
and May 3, 2010

Preface to the First Edition

The soul of every mathematician is wrestled for by the Devil of Abstract Algebra and the Angel of Topology.

—Hermann Weyl

From the left: Angel of Topology, Alexander Soifer and Vladimir Boltyanski while working on the draft of first edition of this book. Colorado Springs, June 1990. ("Angel" is actually a marble by the Italian sculptor Ada Cipriani, born 1904, that commands my living room.)

Mathematics is frequently divided into elementary and higher mathematics, just as literature is divided into children's and grown-up's literature. We do not quite agree with this "discrimination" based on age. It would be more productive if we were to divide both mathematics and literature into good and not-so-good. Accordingly, we decided to make some "grown-up mathematics" available to young mathematicians and their teachers.

Joy of creation, depth and beauty of ideas, flight of fantasy, and unexpected elegance of reasoning, which are so characteristic of mathematics, often remain outside of textbooks. Only popular books and mathematical olympiads enable students to peek into the "Wonderland of Mathematics".

The intellectual eye of a child opens to new and unusual problems, shining summits of magnificent new theories, unexpected "bridges" connecting these summits with each other and uniting them in one wonderland. And most importantly, there is the joy of creating the opportunity to discover new, unexplored corners in the world of mathematics and then to notice with surprise that there is an unexpected path from these behind-the-cloud peaks that opened up the intellectual eye to the real world of things and happenings, to creating new machines and instruments, to solutions of life's problems — problems that previously seemed hopeless.

The main problem of popular literature is in opening, for an interested student, the mysterious world of contemporary mathematics, and, moreover, in bringing him to the forefront of this battle where he will be able join with prominent scientists in the fight to bring out unknown new facts, ideas, and methods. Let these discoveries at first be small, but let them be. The only way for that is to work, to solve problems, and to overcome difficulties. We offer to you, our reader, not easy entertainment, but work and activity that calls forward and inspires.

The authors of this book both love geometry. It is a remarkable region of the Wonderland of Mathematics. Moreover, geometry is not only an important part of the science; for us (as for the majority of mathematicians) geometry is a unique perception of the world that shines a bright light on other areas of mathematics.

It often happens that while solving problems from algebra, analysis, logic or combinatorics, a mathematician draws in front of his intellectual eye a geometric picture that becomes more and more clear, detailed, and understandable—and suddenly geometric insight clears up completely an algebraic or combinatorial problem. The mathematician sits down at the table and writes dozens of formulas, integrals, and equations leading to the goal, the solution to a new problem, a problem that is not at all geometric in its context. Without geometric ideas and representations, the mathematician would have (searched) long and painfully for a solution, like a blind kitten losing the road, getting into dead ends, or senselessly wandering in circles.

We would be very happy if this book gives the reader the opportunity to broaden a little his geometric horizons, and to believe in the magical strength of geometric ideas in the unending world of mathematics. As for combinatorics, probably no mathematician today can formulate precisely what combinatorics is and what problems and methods should be considered combinatorial. But more and more mathematicians invest their efforts in the development of new combinatorial directions in mathematics.

We invite young mathematicians to join this movement, this journey to discover the "New World." The joy of creating, stubborn hardwork, and the ability to cheer up if everything does not come out right from the beginning are the main tools in this journey in which there are no losers, but only winners.

And now a few words about us and how this book was created.

Who are we? A Soviet and an American mathematicians. We got together in beautiful Colorado Springs and in a few concentrated weeks of long hours of writing, discussing, and problem solving each day and night, we produced the first rough draft of this book. It then took the second author eight months of editing, proofing, and adding new material to bring this book to its final form.

This book discusses a few areas of combinatorial mathematics that have something in common. That something is a geometric flavor that we believe adds a visual appeal and distinctive beauty to mathematical reasoning. All four chapters, Tiling, Proofs of Existence, Graphs, and Combinatorial Geometry, show that there is no border between the problems of mathematical olympiads and research problems of mathematics. They introduce our young reader to some exciting ideas and concepts that are not easily available to them from other sources.

We hope life will enable us to continue our joint efforts in the future. We hope to produce a whole library of books for young and talented mathematicians.

We are grateful to Philip Engel, Paul Erdős, Martin Gardner, Branko Grünbaum, and Cecil Rousseau for being the first readers of our manuscript and providing us with valuable feedback. We are honored that Paul Erdős, Branko Grünbaum, and Cecil Rousseau have written introductions for this book.

Our friend and secretary, Lynn Scott, had to put up with two handwritings, one written all over the other (that is what comes out of joint efforts!). Thank you, Lynn!

Lilia Pashkova-Boltyanski took good care of our diet as we worked long hours on the book. Maya Soifer provided valuable help in producing illustrations. Thank you, wives!

We want a dialogue with you, our reader. Beautiful solutions, new problems, or whatever comes from your reading of our book interests us a great deal. Please share it all with us!

Alexander Soifer
University of Colorado
P.O. Box 7150
Colorado Springs, CO 80933
United States of America

Vladimir Boltyanski
National Research Institute of System Research
9pr. 60-letia Oktyabrya
117312 Moscow
Russia
January 1991

Part I
Original Etudes

Chapter 1
Tiling a Checker Rectangle

1 Introduction

Imagine you have an $m \times n$ rectangle R and lots of dominoes (a *domino* is a 1×2 rectangle). It is easy to find the conditions under which R can be *tiled* by dominoes, i.e., covered by dominoes, without any dominoes overlapping or sticking out over the boundary of R. Indeed, R can be tiled by dominoes if and only if mn is even (prove it!).

The problem becomes a bit more difficult if we want to tile the same rectangle with exactly two monominoes (a *monomino* is a 1×1 square) and many dominoes (Figure 1.1). Where can these two monominoes be placed?

In order to answer this question we color the rectangle R in a chessboard fashion in two colors (Figure 1.2).

This coloring has a very nice property: regardless of how a domino is placed on the board, horizontally or vertically, it will cover exactly one square of each color (Figure 1.2). Therefore, the two monominoes *must* cover squares of different colors.

The famous mathematician Ralph E. Gomory found (and apparently never published himself) a beautiful way to prove the converse, that no matter where the two monominoes are placed on the $m \times n$ board (where mn is even and both m and n are greater than 1), as long as they are on different colors, the rest of the board can be tiled by dominoes.

Here is his proof. Supposing n to be even (note at least one of the numbers m, n must be even), Gomory created a labyrinth out of the board (Figure 1.3).

A. Soifer, *Geometric Etudes in Combinatorial Mathematics*,
DOI 10.1007/978-0-387-75470-3_1, © Alexander Soifer, 2010

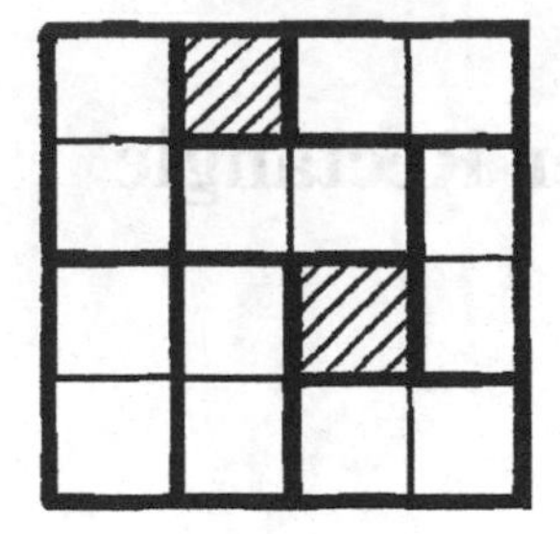

Fig. 1.1

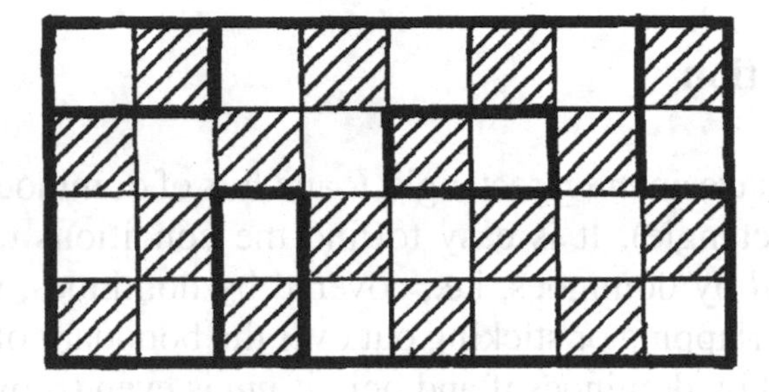

Fig. 1.2 Property of chessboard coloring

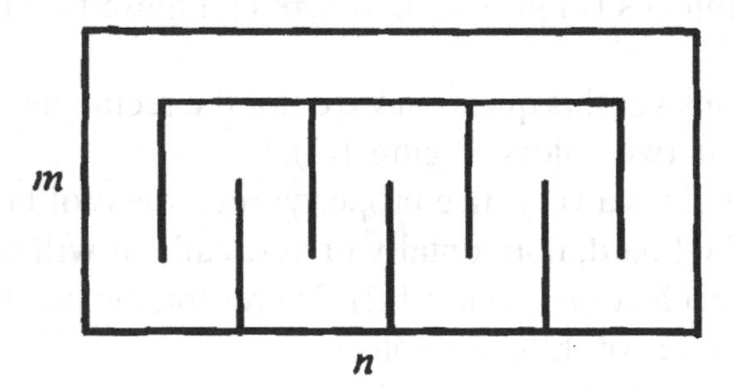

Fig. 1.3 Gomory's maze

As you walk through this labyrinth, black and white squares alternate. Now let us cut the board along the walls of the labyrinth to get a checkered ring with alternating black and white squares (Figure 1.4).

It is clear that no matter what black and white squares we cover by monominoes, the number of squares between the monominoes (in both paths connecting them) must be even and therefore the rest of the ring can be tiled by dominoes.

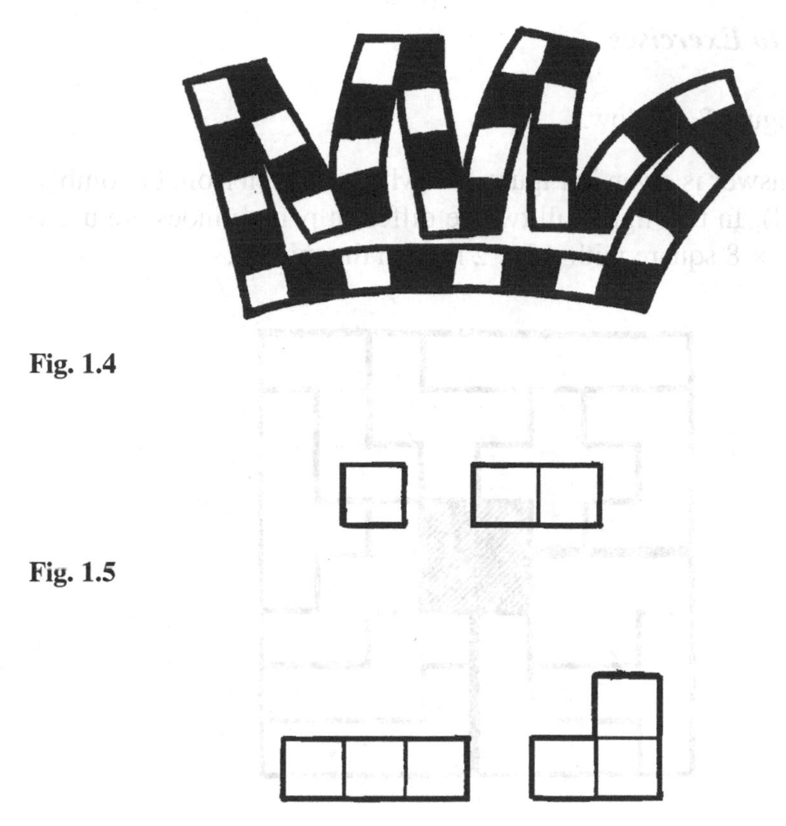

Fig. 1.4

Fig. 1.5

Fig. 1.6 Trominoes

Solomon W. Golomb in his pioneering 1954 paper [Go1] generalized the notion of domino (see also [Go2]). He named the *n-omino* a figure made up of n squares of a checkerboard that are rook-connected, i.e., you can move from any square of the n-omino to any other by horizontal and vertical moves within the figure itself.

You are already familiar with the monomino and domino (Figure 1.5). In Figure 1.6 you can find all shapes of *trominoes* (3-minoes).

Exercise 1.1. Find all shapes of tetrominoes (4-minoes).

Exercise 1.2. Find all shapes of pentominoes (5-minoes).

Solutions to Exercises

1.1. See Figure 3.1 below.

1.2. The answer is given in Figure 1.7 (which is taken from Golomb's book [Go2]). In this figure all twelve different pentominoes are used to tile the 8×8 square with a 2×2 hole in the middle.

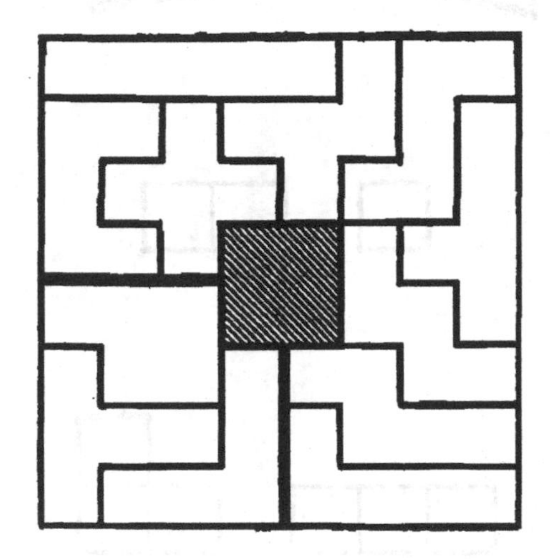

Fig. 1.7

2 Tiling Rectangles by Trominoes

Problems of tiling figures with tiles of an indicated shape (or cutting figures into parts of a given form) form a very interesting area of combinatorial geometry. It includes many fascinating exercises and research problems. Some of them we will consider in this section.

Here is an *L-tromino* (Figure 2.1). It is composed of three unit squares. (Note that there is one other tromino, the *linear tromino*, connecting three squares in a row. We will use this tromino in Section 4.) In this section we will address the following problem: *How to tile a rectangle with L-trominoes.* The first question is how to tile the 2×3 rectangle using L-trominoes. A solution is trivial (Figure 2.2).

Fig. 2.1 L-tromino

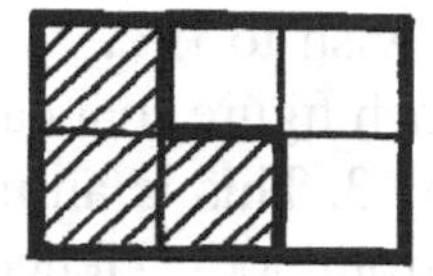

Fig. 2.2

We now offer a sequence of exercises. You are invited to find solutions. After several attempts (successful or not) we recommend you read our solutions at the end of the section.

Exercise 2.1. Is it possible to tile a 4×5 rectangle with L-trominoes?

Exercise 2.2. Find the smallest square that can be tiled with L-trominoes.

Exercise 2.3. Find all integers b such that a $2 \times b$ rectangle can be tiled using L-trominoes.

Exercise 2.4. Find all integers b such that a $3 \times b$ rectangle can be tiled by L-trominoes.

Combining solutions to Exercises 2.3 and 2.4, we obtain the following result.

Theorem 2.1. *Let a, b be integers such that $2 \le a \le 3$ and $a \le b$. The $a \times b$ rectangle can be tiled by L-trominoes if and only if ab is divisible by 6.*

Equivalently: the $a \times b$ rectangle with $2 \le a \le 3$ and $a \le b$ can be tiled by L-trominoes if and only if one of the numbers a, b is divisible by 2 and the other is divisible by 3.

Now a new direction of research is clear: we will consider $a \times b$ rectangles in cases when $4 \le a \le b$. But first we would like to share a couple of "philosophical" observations.

Suppose we are trying to solve a combinatorial problem. For example, perhaps we are trying to decide whether it is possible to tile a given figure F by L-trominoes. Let us assume further that we expect the answer to be "no". How do we find out? If some attempts to tile F with L-trominoes were unsuccessful, we may not yet give an answer. Indeed, maybe the next attempt will be more successful. So, how does one establish that F cannot be tiled by L-trominoes? There are, in general, two ways. The first way consists of using an *invariant* property of the figures we wish to tile. Note that the area of the L-tromino is equal to 3. So, each figure that can be tiled by L-trominoes has to have area divisible by 3. This is an invariant, general property of all figures tileable by L-trominoes. Hence, if the figure F does not possess this property (that is, if its area is not divisible by 3), then F cannot be tiled by L-trominoes. Such an approach is used in the solution of Exercise 2.1 (and others). Another way to get the answer "no" consists of looking at all possibilities. Such an approach is used in the solution of Exercise 2.2 (see Figure 2.4). These two approaches to getting the answer "no" are most common.

Let us now suppose that we expect to get the answer "yes" (to the question of whether the figure F can be tiled by L-trominoes). In order to get this answer, we need to construct a concrete tiling of F by L-trominoes (or an algorithm for obtaining such a tiling).

You may be wondering what we should do if we do not know what answer to expect. In this case we need to aid our intuition by *experimenting*, i.e., by considering and solving relatively small and simple cases of the problem.

At this point you, our reader, are well armed to prove Theorem 2.2. We created the following sequence of exercises to make your climbing to the summit of Theorem 2.2 easier.

Exercise 2.5. Prove that a 5×6 rectangle can be tiled by L-trominoes.

Exercise 2.6. Prove that a 5×9 rectangle can be tiled by L-trominoes.

Exercise 2.7. Prove that a 9×9 square can be tiled by L-trominoes.

Exercise 2.8. Prove that if $b > 5$ and b is divisible by 3, then a $5\times b$ rectangle can be tiled by L-trominoes.

Exercise 2.9. Prove that if b is divisible by 3 and an $a\times b$ rectangle can be tiled by L-trominoes, then an $(a+2)\times b$ rectangle can also be tiled by L-trominoes.

Exercise 2.10. Let integers a, b satisfy the conditions $a \geq 4$, $b \geq 5$, and b divisible by 3. Decide whether an $a \times b$ rectangle can be tiled by L-trominoes.

With the help of the solution to Exercise 2.10, we can obtain the following result:

Theorem 2.2. *Let F be the $a \times b$ rectangle where $a \geq 4$, $b \geq 4$. Then F can be tiled by L-trominoes if and only if the product ab is divisible by 3.*

Finally, we can combine statements of Theorems 2.1 and 2.2:

Theorem 2.3. *Let a, b be integers such that $2 \leq a \leq b$. The $a \times b$ rectangle can be tiled by L-trominoes if and only if one of the following cases holds:*

i) $a = 3$ and b is even;
ii) $a \neq 3$ and ab is divisible by 3.

Solutions to Exercises

2.1. The area of an L-tromino (see Figure 1.6) is equal to 3. The area of the 4×5 rectangle is equal to 20. Hence, in order to tile this rectangle by L-trominoes, it is necessary to use $\frac{20}{3}$ L-trominoes, which is impossible.

Generally, if an $a \times b$ rectangle can be tiled by L-trominoes, then ab is divisible by 3. This is a necessary condition for the existence of such a tiling. ■

2.2. Due to Exercise 2.1, if an $a \times a$ square can be tiled by L-trominoes, then a is divisible by 3. The 6×6 square can be decomposed into 2×3 blocks (Figure 2.3), and consequently, the 6×6 square can be tiled by L-trominoes (see Figure 2.2). So, it remains to be seen whether the 3×3 square can be tiled by L-trominoes. The answer is "no." Indeed, all the ways to cover the upper-left corner are shown in Figure 2.4.

In Figure 2.4(a), it is impossible to cover the upper right corner. In Figure 2.4(b), it is impossible to cover the lower left corner. Finally, in Figure 2.4(c), the only ways to cover the upper right corner is shown

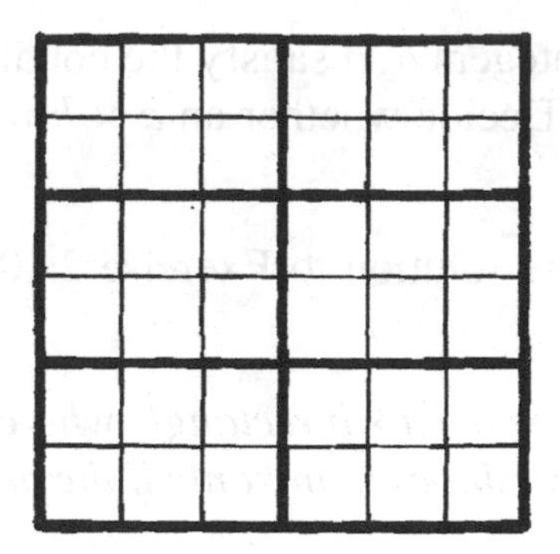

Fig. 2.3

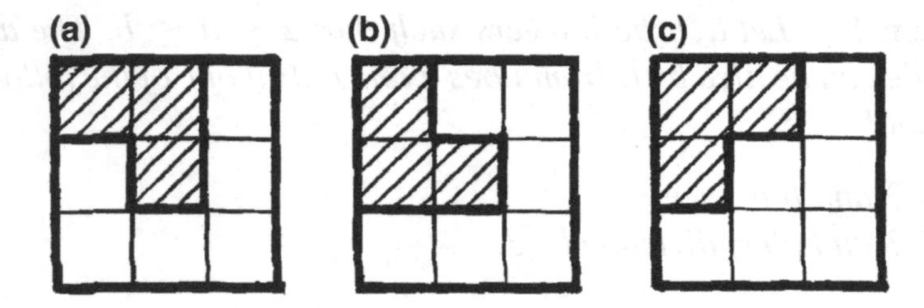

Fig. 2.4

Fig. 2.5

in Figure 2.5. But then the three bottom squares remain uncovered. Thus the investigation of all possibilities shows that the 3×3 square cannot be tiled by L-trominoes. Consequently, the 6×6 square is the smallest tileable square. ■

2.3. If b is divisible by 3, then a $2 \times b$ rectangle can be decomposed into 2×3 blocks (see Figure 2.6) and, consequently, this rectangle can be tiled by L-trominoes. The reasoning in the solution to Exercise 2.1 shows that there are no other possibilities. So, the $2 \times b$ rectangle can be tiled by L-trominoes if and only if b is divisible by 3. ■

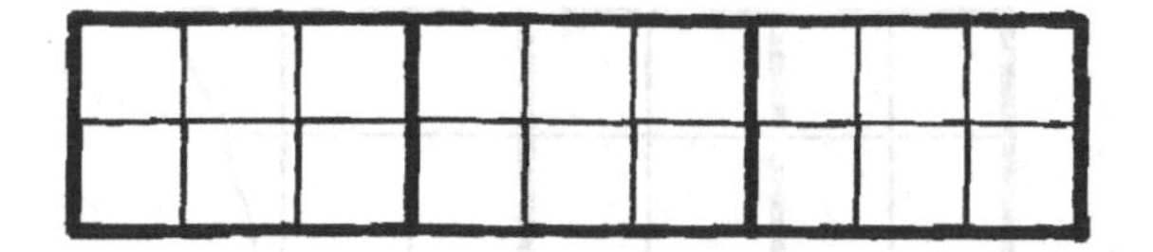

Fig. 2.6

2.4. Let us consider a $3 \times b$ rectangle. All the ways to cover the upper left corner are shown in Figure 2.7. But in Figure 2.7(c), it is impossible to cover the lower left corner. As for the possibilities in Figures 2.7(a) and 2.7(b), the lower left corner can be covered uniquely. This is shown in Figure 2.8.

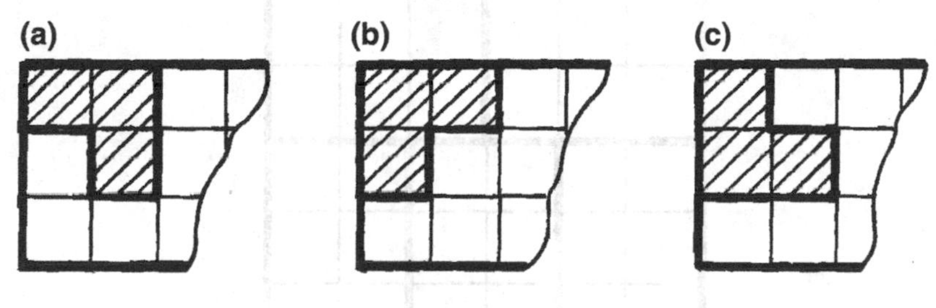

Fig. 2.7

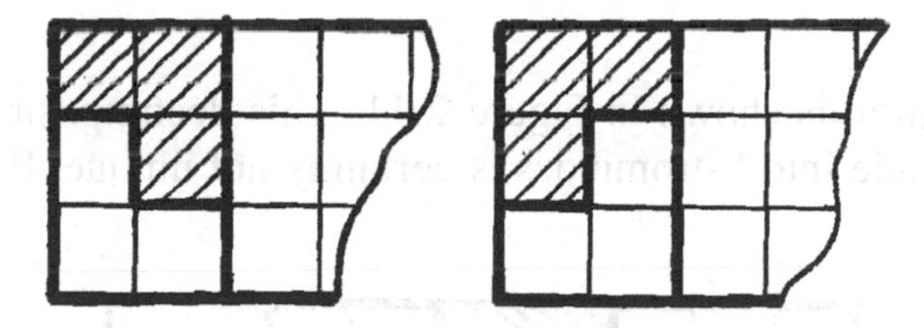

Fig. 2.8

In either case, we have the 3×2 block that is situated on the left side of the $3 \times b$ rectangle and covered by two L-trominoes. In the same manner we obtain the next 3×2 block covered by two L-trominoes (Figure 2.9) and so on.

This reasoning shows that if the $3 \times b$ rectangle can be tiled by L-trominoes, then b is even. This condition is not only necessary but also sufficient: if b is even, then the $3 \times b$ rectangle can be decomposed into 3×2 blocks. ■

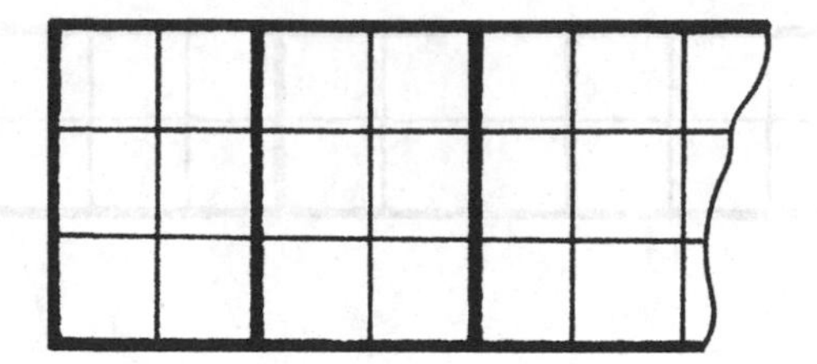

Fig. 2.9

2.5. The 5×6 rectangle can be decomposed into 2×3 blocks (Figure 2.10). Thus the rectangle can be tiled by L-trominoes. ■

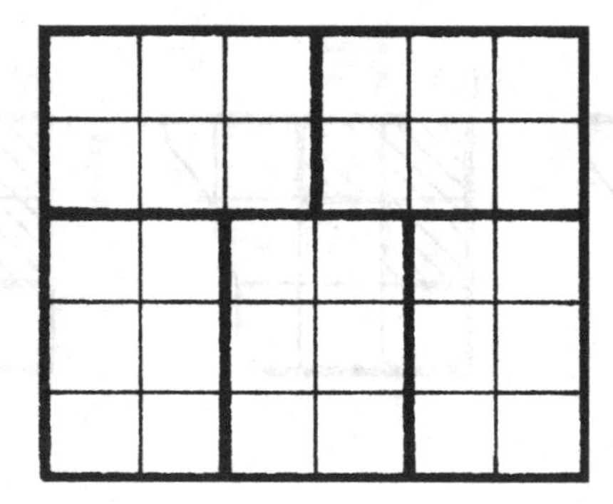

Fig. 2.10

2.6. A solution is shown in Figure 2.11. This decomposition of the 5×9 rectangle into L-trominoes is certainly not unique. Please note

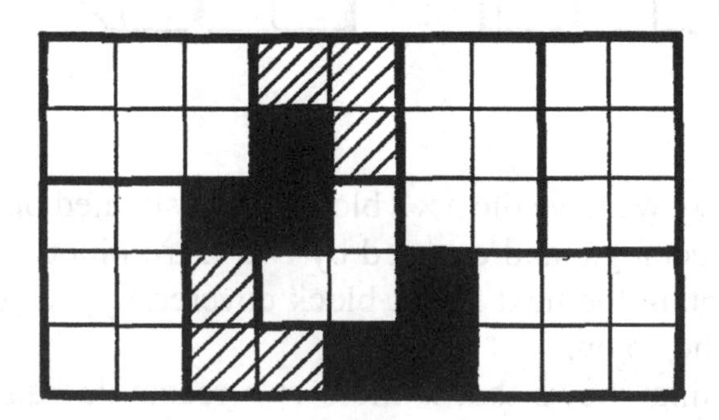

Fig. 2.11 Tiling the 5×5 rectangle with L-trominoes

that in this exercise we have the $a \times b$ rectangle with $a = 5, b = 9$, so ab is not divisible by 6 (only by 3), yet the rectangle is tileable by L-trominoes. ■

2.7. The 9×9 square can be decomposed into two rectangles that are 4×9 and 5×9 (Figure 2.12). Each of these rectangles can be tiled by L-trominoes (according to the results of Exercises 2.3 and 2.6). Thus the 9×9 square can be tiled by L-trominoes. ■

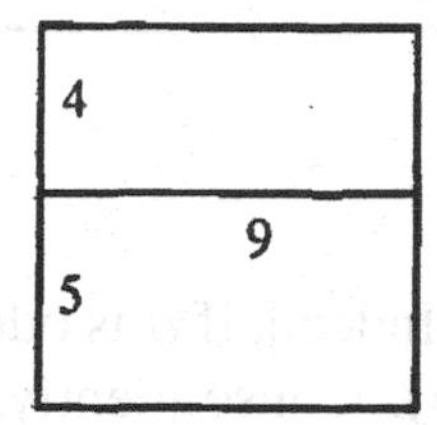

Fig. 2.12

2.8. Each $5 \times b$ rectangle where b is divisible by 3 and $b \geq 6$ can be decomposed into blocks 5×6 and 5×9. Indeed, if b is divisible by 6, then it suffices to use only 5×6 blocks (Figure 2.13). And if b is not divisible by 6, then it is necessary to use one 5×9 block (see Figure 2.14). ■

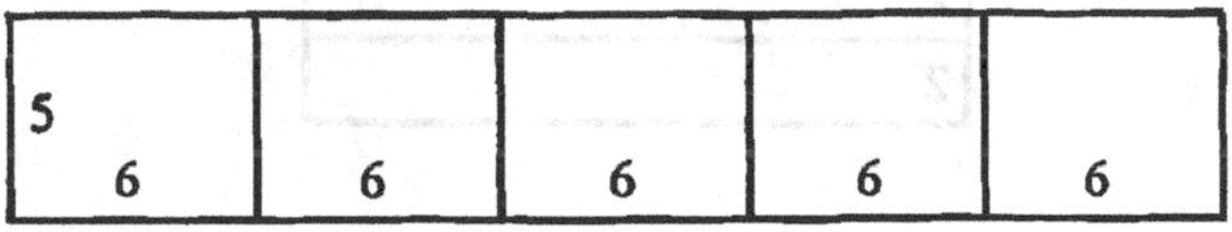

Fig. 2.13

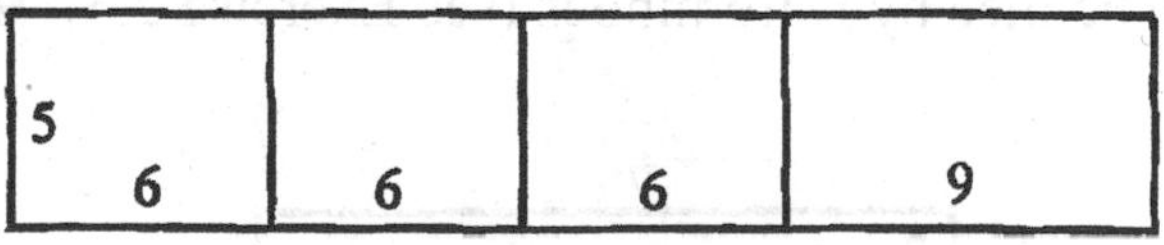

Fig. 2.14

2.9. A solution is sketched in Figure 2.15. Since b is divisible by 3, a $2 \times b$ rectangle can be decomposed into 2×3 blocks. So a $2 \times b$ rectangle can be tiled by L-trominoes. Therefore an $(a+2)\times b$ rectangle can also be tiled by L-trominoes. ■

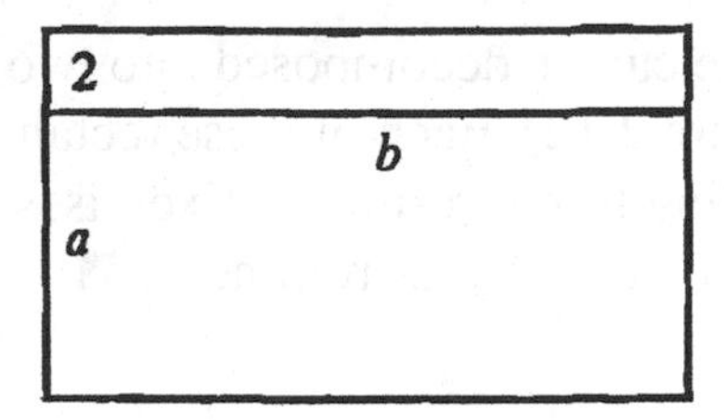

Fig. 2.15

2.10. The answer is "yes." Indeed, if a is odd, then $a = 5+2k$ where $k \geq 0$ (recall that $a \geq 5$). Consequently, an $a \times b$ rectangle can be decomposed into one $5 \times b$ block and a number of $2 \times b$ blocks (Figure 2.16). Thus an $a \times b$ rectangle can be tiled by L-trominoes (see Exercises 2.8 and 2.9).

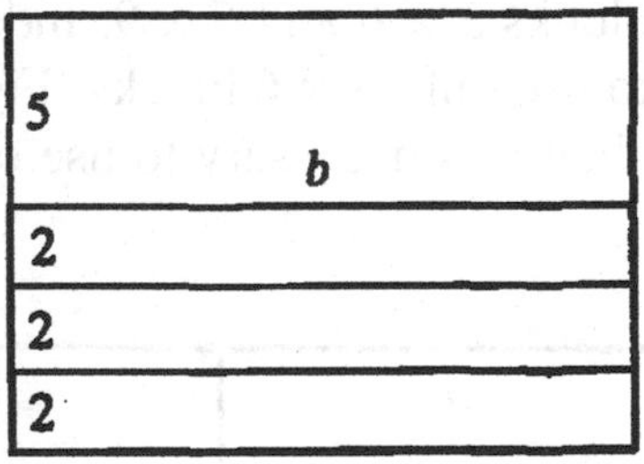

5
b
2
2
2

Fig. 2.16

If a is even, then an $a \times b$ rectangle can be decomposed into a number of $2\times b$ blocks (Figure 2.17) and, consequently, an $a \times b$ rectangle can be tiled by L-trominoes (see Exercise 2.3). ■

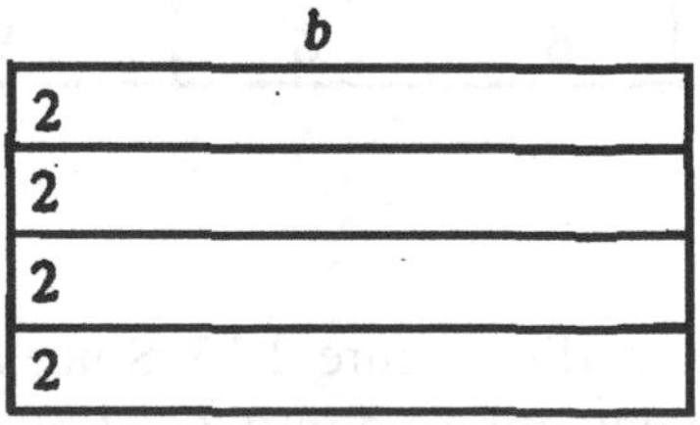

b
2
2
2
2

Fig. 2.17

3 Tetrominoes and Chromatic Reasoning

There exist five types of *tetrominoes* (Figure 3.1). We will call them O-, Z-, L-, T-, and I-tetrominoes. In this section we consider the problem of tiling a rectangle by O-, Z-, L-, and T-tetrominoes. For I-tetrominoes, the problem will be generalized and solved in the next section. In order to solve these problems we will often use "color" reasoning that was mentioned in Section 1.

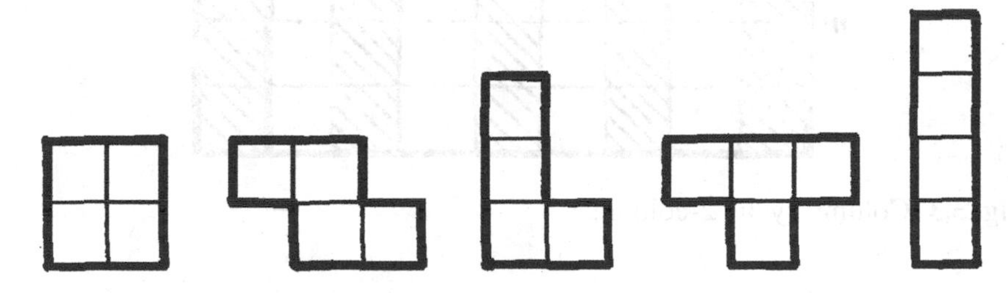

Fig. 3.1 Tetrominoes

Exercise 3.1. Prove that the $m \times n$ rectangle can be tiled with O-tetrominoes if and only if m, n are even.

Exercise 3.2. Prove that no rectangle can be tiled with Z-tetrominoes.

The above exercises do not require the use of "color" reasoning. But such reasoning will be very useful for other types of tetrominoes.

Exercise 3.3. Prove that if all squares of an $a \times b$ chess-board are colored in two colors black and white (Figure 3.2), then the difference between the number of black squares and the number of white squares is equal to either 0 or ± 1.

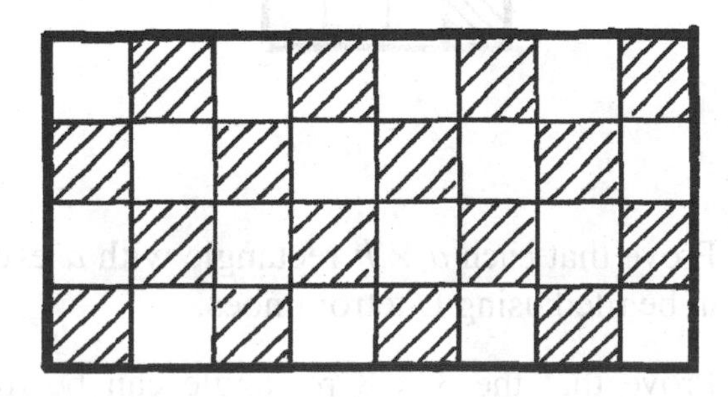

Fig. 3.2 Chessboard coloring

Exercise 3.4. Prove that if an $m \times n$ rectangle is colored in two colors, black and white, by columns (Figure 3.3), then the difference between the number of black squares and the number of white squares is equal to either 0 or $\pm m$.

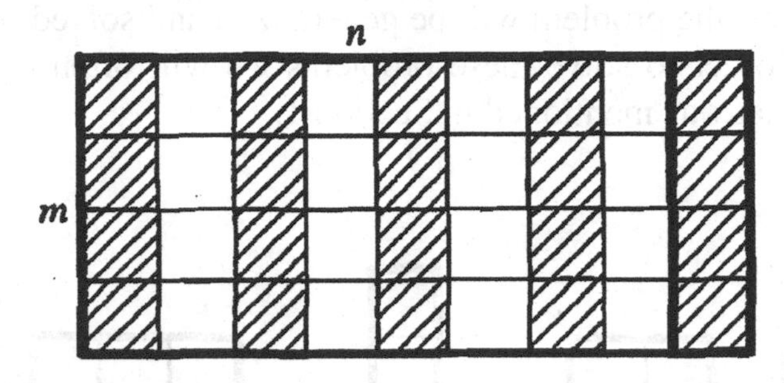

Fig. 3.3 Column cyclic 2-coloring

Exercise 3.5. Prove that if an $a \times b$ rectangle is tiled with L-tetrominoes, then the number of tiles is even.

This exercise shows that the area of a rectangle tiled by L-tetrominoes must be divisible by 8.

It is obvious that the 2×4 rectangle can be tiled by two L-tetrominoes (Figure 3.4). The following problem arises naturally: *What other rectangles can be tiled by tetrominoes of this type?* Can you solve this problem on your own? If not, try to solve the following exercises to pave the way for a successful train of thought.

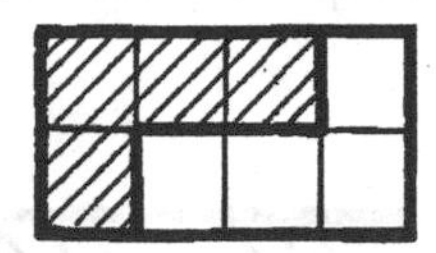

Fig. 3.4

Exercise 3.6. Prove that each $a \times b$ rectangle with a even and b divisible by 4, can be tiled using L-tetrominoes.

Exercise 3.7. Prove that the 3×8 rectangle can be tiled with L-tetrominoes.

Exercise 3.8. ([GK]) Prove that an $a \times b$ rectangle with $a \geq 2$ and $b \geq 2$ can be tiled with L-tetrominoes if and only if mn is divisible by 8.

The above exercise gives a complete solution tiling a rectangle with L-tetrominoes. Solving a similar problem for T-tetrominoes is much more difficult. We leave the first steps to the reader.

Exercise 3.9. Prove that if an $m \times n$ rectangle is tiled by T-tetrominoes, then the number of tiles is even.

Thus, the area of a rectangle tiled with T-tetrominoes is divisible by 8.

Exercise 3.10. Prove that the 4×4 square can be tiled using T-tetrominoes.

Exercise 3.11. Prove that an $n \times n$ square can be tiled using T-tetrominoes if and only if n is divisible by 4.

It is clear that an $m \times n$ rectangle with each of the numbers m, n divisible by 4 can be tiled by T-tetrominoes (Figure 3.5; Exercise 3.10). This condition sufficient, for the tileability with T-tetrominoes, is also necessary. A clever proof of this fact was found by D. W. Walkup of Boeing Scientific Research Laboratories in 1965. Read it in the *Monthly!*

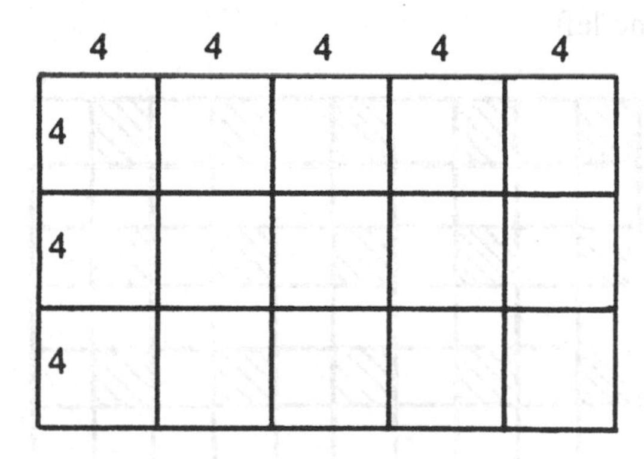

Fig. 3.5 Tiling of the 3×8 rectangle with L-tetrominoes

Theorem 3.1. *(D. W. Walkup [Wa]) If an $m \times n$ rectangle can be tiled with T-tetrominoes, then m and n are multiples of 4.*

Solutions to Exercises

3.1. If both numbers m, n are even, then the $m \times n$ rectangle can obviously be tiled (Figure 3.6).

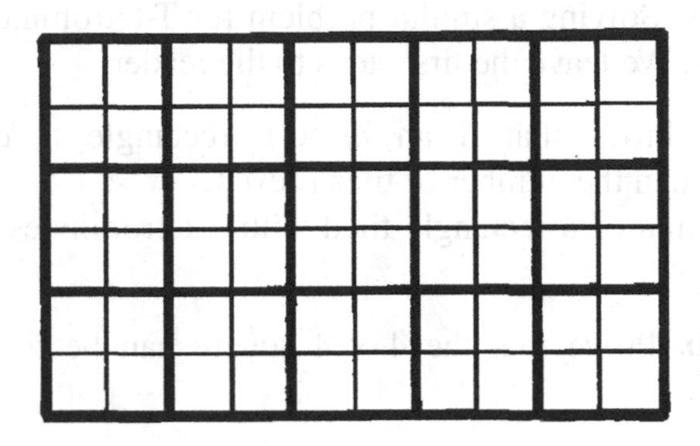

Fig. 3.6

In order to prove the converse, let us consider the coloring shown in Figure 3.7. Each O-tetromino covers exactly *one* black square. Thus, if the rectangle is tiled with O-tetrominoes, then the number of tetrominoes in a row is equal to a, where a is the number of the black squares in the upper row. Consequently, the number of the unit squares in a row is equal to $2a$. Similarly, the number of the unit squares in a column is equal to $2b$, where b is the number of black squares in the first column from the left.

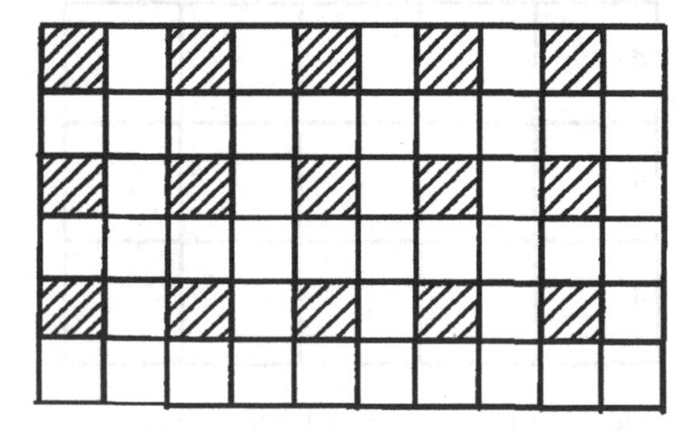

Fig. 3.7

We can solve this exercise without coloring. Indeed, let a rectangle tiled with O-tetrominoes be $a \times b$ where the side of the tetromino is taken as the unit of measure. If we use to the side of a unit square as the measuring unit, then the rectangle would have size $2a \times 2b$. ■

3.2. There are two ways to cover the upper left square with Z-tetrominoes (Figure 3.8). The two ways are symmetric, so it suffices to consider only the first one. Then, there is only one way to cover the third square in the upper row (Figure 3.9). Further, there is only one way to cover the fifth square in the upper row (Figure 3.10) and so on.

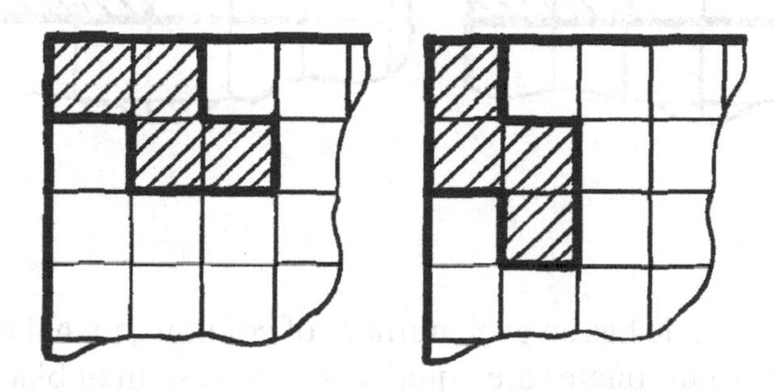

Fig. 3.8

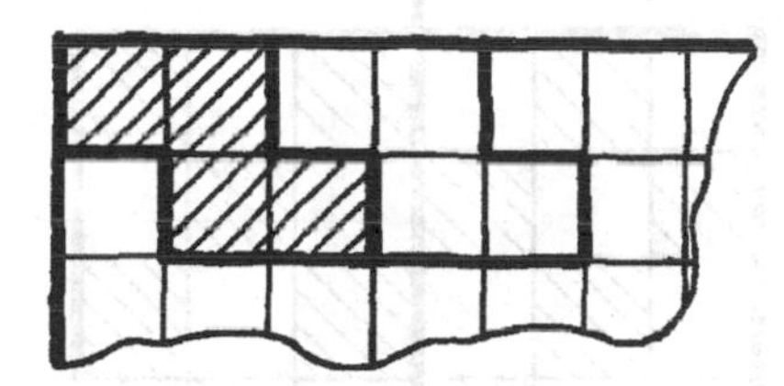

Fig. 3.9

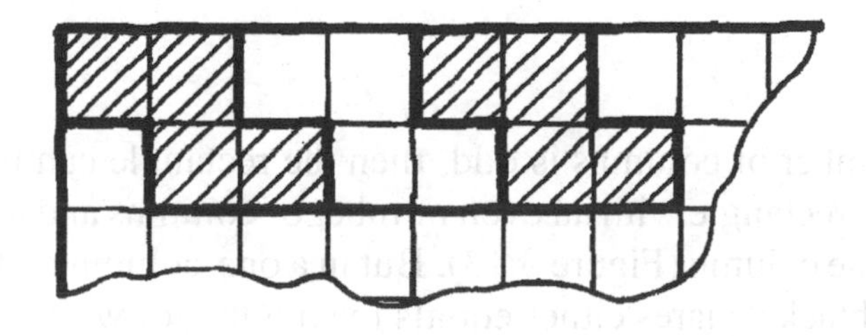

Fig. 3.10

When we get to the right side of the rectangle (Figure 3.11), it is impossible to cover the upper right corner (and stay inside the boundaries of the rectangle). Thus it is impossible to tile any rectangle with Z-tetrominoes. ■

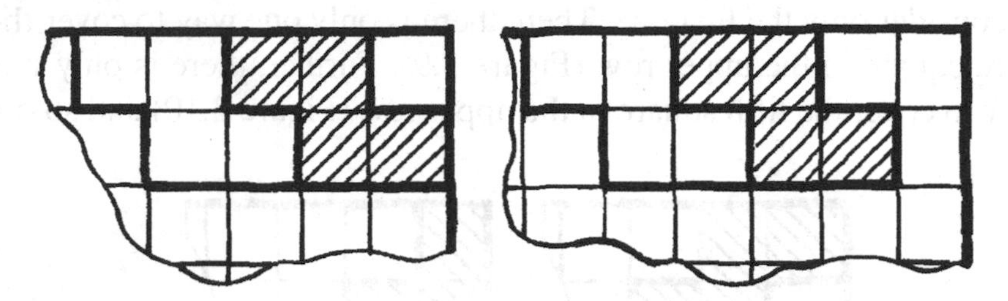

Fig. 3.11

3.3. If the rectangle has an even number of columns, then the numbers of black and white squares are equal since the system of black squares is symmetric to the system of white squares with respect to the midline (Figure 3.12).

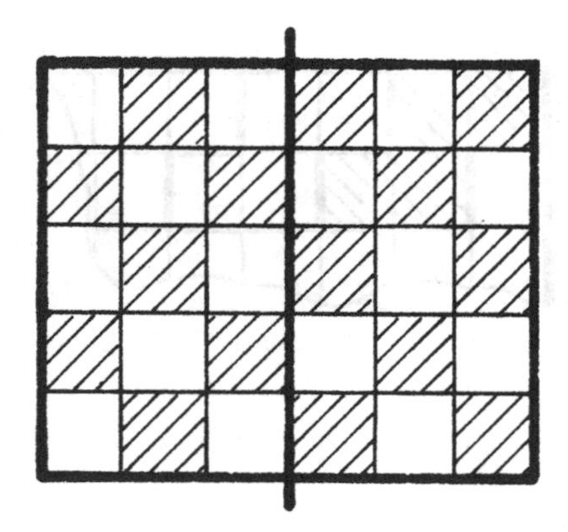

Fig. 3.12

If the number of columns is odd, then the rectangle can be decomposed into a rectangle with an even number of columns and a rectangle with only one column (Figure 3.13). But in a one-column rectangle the number of black squares either equals the number of white squares or differs from that number by one. ■

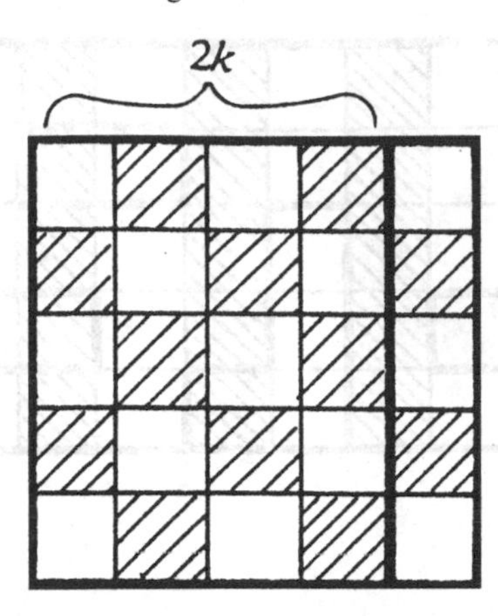

Fig. 3.13

3.4. If the number of columns is even, then the number of black squares equals the number of white squares. If the number of columns is odd, then there is one "extra" column. In this case the difference between the number of black and the number of white squares is equal to $\pm m$ (Figure 3.3). ■

3.5. Assume that the $a \times b$ rectangle can be tiled with L-tetrominoes. Since the area of a L-tetromino is equal to 4, $ab = 4k$, where k is the number of L-tetrominoes. Consequently, ab is divisible by 4, so at least one of the numbers a, b is even. Without loss of generality we can assume that b is even, that is, the number of columns in the $a \times b$ rectangle is even.

Let us color the rectangle by alternating black and white columns. Any L-tetromino placed on the colored rectangle covers three unit squares of one color and one unit square of the other color (Figure 3.14). So we have "black" L-tetrominoes (three black squares and one white square) and "white" L-tetrominoes (three white squares and one black square). If the number p of "black" L-tetrominoes is not equal to the number q of "white" L-tetrominoes, then the number of black and white squares in the rectangle are equal. But they are equal, since n is even; therefore, $p = q$, that is, the number $p + q$ of all L-tetrominoes is even. Therefore, the area mn of the rectangle is divisible by 8. ■

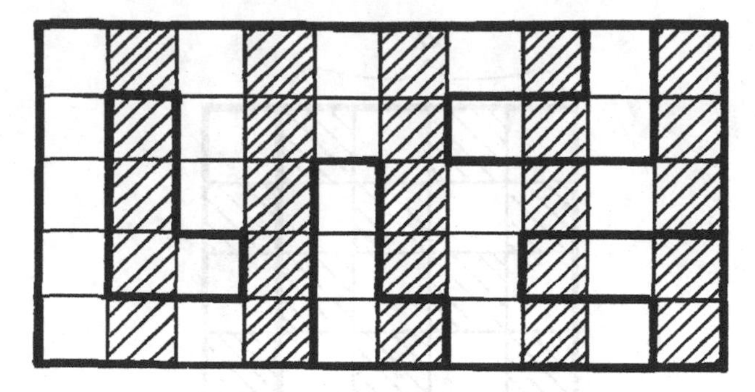

Fig. 3.14

3.6. An $a \times b$ rectangle can be decomposed into 2×4 blocks, and consequently, it can be tiled by L-tetrominoes (see Figure 3.4). ■

3.7. A solution is shown in Figure 3.15. This can be said to be "essentially" the only solution. What do you think this uniqueness means? ■

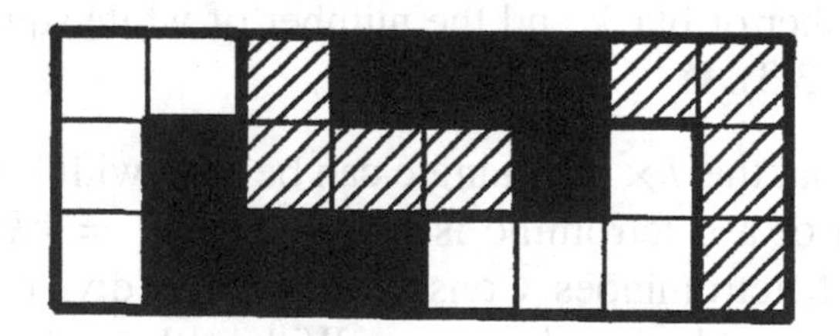

Fig. 3.15 Tiling of the 3×8 rectangle with L-tetrominoes

3.8. Divisibility by 8 is necessary. This follows from the solution to Exercise 3.5. Now we prove that it is also sufficient.

Let M be an $a \times b$ rectangle such that the product mn is divisible by 8. If both of the numbers a, b are even (and one of them is divisible by 4), then M can be tiled by L-tetrominoes (see Exercise 3.6). If one of the numbers a, b (say a) is odd, and, consequently, the other is divisible by 8, then M can be decomposed into $a \times 8$ blocks (Figure 3.16). Further, since $a \geq 2$ is odd, then it is possible to decompose the $a \times 8$ block into a 3×8 block and an $(a - 3) \times 8$ block where $a - 3$ is even (Figure 3.17). Consequently (see Exercises 3.6 and 3.7) the $a \times 8$ block can be tiled by L-tetrominoes. Thus, the $a \times b$ rectangle can be tiled as well. ■

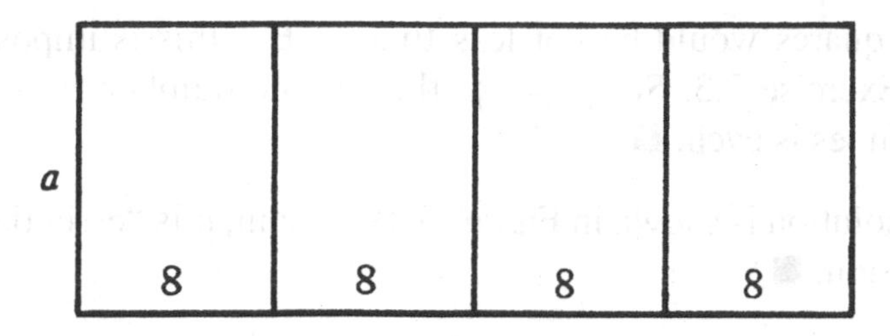

Fig. 3.16

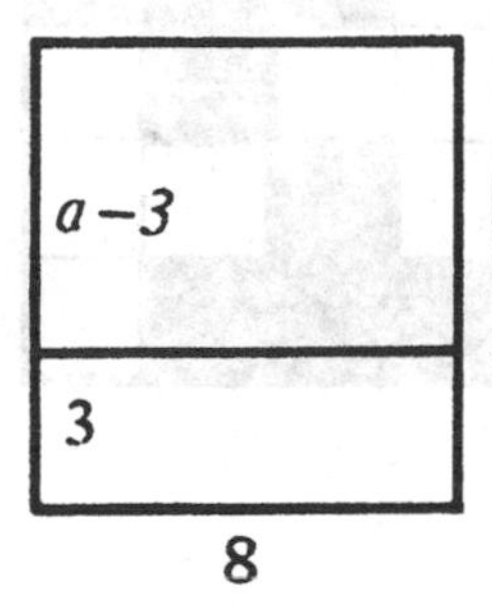

Fig. 3.17

3.9. Consider the "chessboard" coloring of the rectangle. It can easily be shown that each T-tetromino placed on the rectangle covers three unit squares of one color and one unit square of the other color (Figure 3.18). So we have "black" T-tetrominoes (three black squares and one white square) and "white" T-tetrominoes (three white squares and one black square). If the number p of "black" T-tetrominoes were not equal to the number q of "white" T-tetrominoes, then the difference between the number of "black" squares and the number of

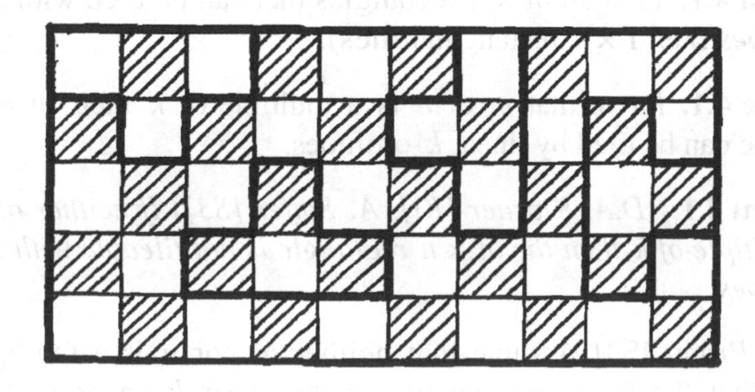

Fig. 3.18

"white" squares would be not less than 2. But this is impossible in light of Exercise 3.3. So, $p = q$, that is, the number $p + q$ of all T-tetrominoes is even. ■

3.10. A solution is shown in Figure 3.19. Again, it is "essentially" the only solution. ■

Fig. 3.19

3.11. If an $n \times n$ square can be tiled with T-tetrominoes, then n^2 is divisible by 8 (see Exercise 3.8), and, consequently, n is divisible by 4. Conversely, if n is divisible by 4, then the $n \times n$ square can be decomposed into 4×4 blocks and hence the $n \times n$ square can be tiled using T-tetrominoes (see Exercise 3.10). ■

4 Tiling by Linear Polyominoes

In this section we first solve the following problem.

Problem 4.1. Find all $m \times n$ rectangles that can be tiled with ***linear k-ominoes*** (i.e., $1 \times k$ rectangular tiles).

Exercise 4.1. Prove that m or n is a multiple of k then an $m \times n$ rectangle can be tiled by linear k-ominoes.

Theorem 4.1. *(D.A. Klarner [Kl], A. Soifer [S3]) If neither of m, n is a multiple of k then the $m \times n$ rectangle is not tileable with linear k-ominoes.*

First Proof. [S3] Assume that neither m nor n is a multiple of k, but that the board can be tiled with linear k-ominoes. Let us

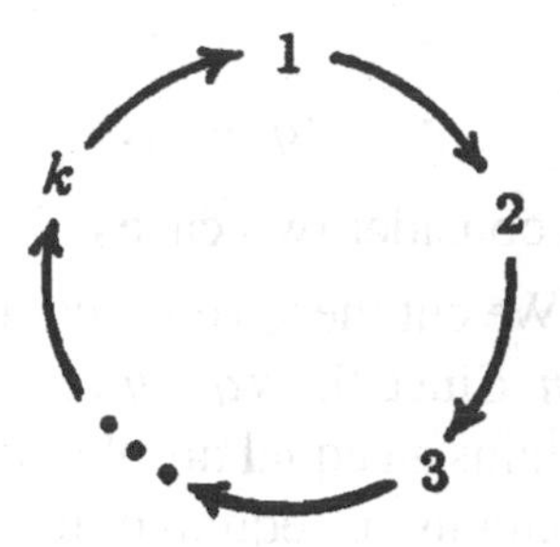

Fig. 4.1

Fig. 4.2 Diagonal cyclic k-coloring for $k = 3$

color the board diagonally in k colors with a cyclic permutation of colors (see Figures 4.1 and 4.2).

This coloring (**diagonal cyclic k-coloring**) has a remarkable property: no matter how a linear k-omino is placed on the colored rectangle, horizontally or vertically, it will cover exactly one square of each of the k colors.

Since by assumption the rectangle can be tiled using linear k-ominoes, and every k-omino covers an equal number of squares of every color, the rectangle must contain an equal number of squares of each of the k colors. It is not difficult to prove (do!) that for any natural numbers m and k there are non-negative integers q and r_1, such that $0 \leq r_1 \leq \frac{k}{2}$ and

$$m = kq + r_1$$

or

$$m = kq - r_1.$$

Accordingly, we will consider two cases:

Case 1: $m = kq + r_1$. We cut the given rectangle into two rectangles that are $kq \times n$ and $r_1 \times n$. Since the $kq \times n$ rectangle can be tiled using linear k-ominoes, it contains an equal number of squares of each color. The given rectangle contains an equal number of squares of every color as well; therefore, the $r_1 \times n$ rectangle has the same property.

Case 2: $m = kq - r_1$. In this case we attach the $r_1 \times n$ rectangle to the given rectangle to obtain a $kq \times n$ rectangle and extend the coloring of the given rectangle to a diagonal cyclic k-coloring of the $kq \times n$ board. The $kq \times n$ rectangle can be tiled by assumption using linear k-ominoes; therefore, it contains an equal number of squares of every color. The given $m \times n$ rectangle contains an equal number of squares of every color; therefore, the $r_1 \times n$ rectangle contains an equal number of squares of every color.

Thus, in both cases we tiled the $r_1 \times n$ rectangle using a diagonal cyclic k-coloring, which contains an equal number of squares of every color. Let us now turn this rectangle 90°, i.e., consider the $n \times r_1$ rectangle, and apply to it all of the above reasoning. As a result we will get an $r_2 \times r_1$ rectangle, where $0 < r_2 \le \frac{k}{2}$, that contains an equal number of squares of every color.

On the other hand, the total number of one-color diagonals in the $r_2 \times r_1$ rectangle is equal to $r_2 + r_1 - 1$, and

$$r_2 + r_1 - 1 \le \frac{k}{2} + \frac{k}{2} - 1 < k.$$

Therefore at least one of the k colors is not present in the $r_2 \times r_1$ rectangle!

This contradiction proves that the $m \times n$ board can be tiled with linear k-ominoes only if m or n is a multiple of k. ■

Second Proof [S3]. *The following notation will be helpful for us: instead of writing "the remainders upon dividing numbers $a_1, a_2, \ldots a_k$ by n are equal" we will simply write*

$$a_1 \equiv a_2 \equiv \ldots \equiv a_k (\text{mod } n)$$

and say that $a_1, a_2, \ldots a_k$ are congruent modulo n.

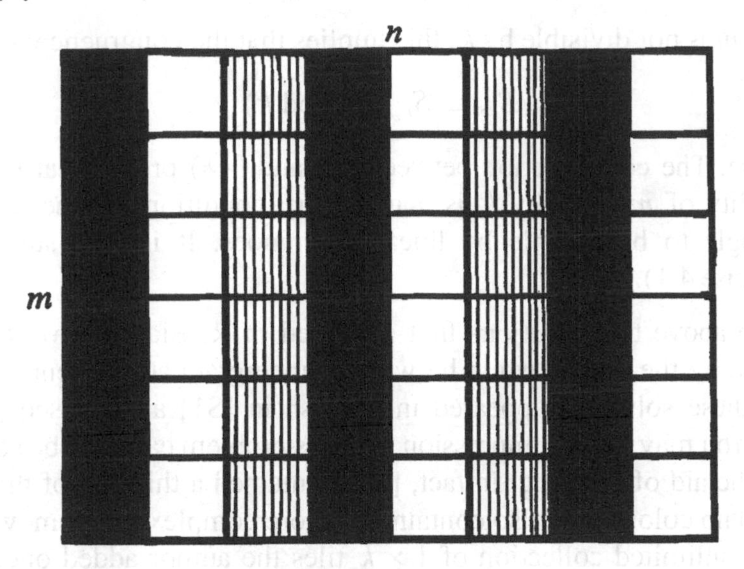

Fig. 4.3 Column cyclic k-coloring for $k = 3$

It is easy to prove (do!) that $a_1 \equiv a_2 \pmod{n}$ if and only if $a_1 - a_2$ is a multiple of n.

Assume that neither of m, n is a multiple of k, i.e.,

$$m = kq_1 + r_1; \quad 0 < r_1 < k,$$
$$n = kq_2 + r_2; \quad 0 < r_2 < k,$$

but the $m \times n$ rectangle is tiled by linear k-ominoes.

Let us color the columns of the rectangle with one of each of the k colors with cyclic permutation of colors (see Figures 4.1 and 4.3) and denote by $S_1, S_2, \ldots, S_n$ the number of squares of the rectangle colored with the 1st, 2nd, ..., k-th colors respectively.

This coloring (**column cyclic *k*-coloring**) has a nice property: if a linear k-omino is placed on the board vertically, it covers k squares of the same color; if it is placed on the board horizontally, it covers exactly one square of every color. By assumption the rectangle is tiled by linear k-ominoes, therefore

$$S_1 \equiv S_2 \equiv \ldots \equiv S_k \pmod{k}. \qquad (*)$$

On the other hand, notice that we have one column more of the r_2-th color than of the $(r_2 + 1)$st color, i.e.,

$$S_{r_2} - S_{r_2+1} = m.$$

Since m is not divisible by k, this implies that the congruence

$$S_{r_2} \equiv S_{r_2+1} \pmod{k} \qquad (**)$$

is false. The contradiction between $(*)$ and $(**)$ proves that the divisibility of m or n by k is a necessary condition for the $m \times n$ rectangle to be tileable by linear k-ominoes. It is also sufficient (Exercise 4.1). ■

The above two solutions first appeared in Russian in *Kvant* [S3] written by the author while he was an undergraduate student. Much later these solutions appeared in English in [S1] and consequently [S9]. You may get the impression that this problem can only be solved with the aid of coloring. In fact, [S3] contained a third proof that required no coloring. It also contained a more complex problem, where to the unlimited collection of $1 \times k$ tiles the author added one, just one, monomino. Let us look at this problem.

Problem 4.2. (A. Soifer [S3]) Which rectangles can be tiled by a combination of one monomino and any number of linear k-ominoes? Where must we place the monomino?

Let us split problem 4.2 into several exercises.

Exercise 4.2. If an $m \times n$ rectangle is tiled with linear k-ominoes and one monomino, then mn–1 is divisible by k.

We need a way to talk about a particular 1×1 square in an $m \times n$ rectangle without using the unit square. We can certainly number all unit squares of the rectangle: $1, 2, \ldots, mn$. But this would be bulky and not very convenient; it would be tricky even to see whether two unit squares are neighbors or not. A much better way is to define a Cartesian coordinate system on our $m \times n$ rectangle (Figure 4.4).

Each unit square is defined now by the ordered pair of integers (x, y), where x and y are the number of the row and column in which the square is located.

Exercise 4.3. Solve Problem 4.2 for $k = 2$.

Now let us solve the main part of Problem 4.2: $k > 2$.

Assume that an $m \times n$ rectangle is tiled using one monomino and linear k-ominoes. Let us color the rectangle in k colors using column

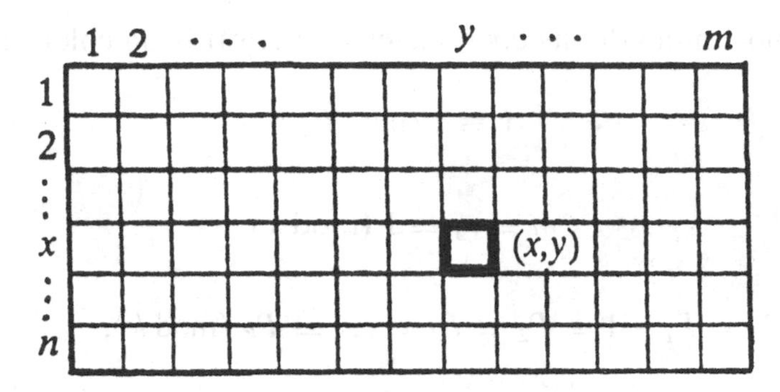

Fig. 4.4 Cartesian coordinates on a checker rectangle

cyclic coloring (see Figures 4.1 and 4.3). As we have already noticed in the solution of Problem 4.1, this coloring has the following property: if a linear k-omino is placed on the rectangle vertically, it will cover k squares of the same color; if it is placed on the rectangle horizontally, it will cover exactly one square of every color. By assumption the $m \times n$ rectangle is tiled by linear k-ominoes and one monomino; therefore,

$$S_1 \equiv S_2 \equiv \ldots \equiv S_k \pmod{k}, \tag{1}$$

where S_i is the number of unit squares of color i ($i = 1, 2, \ldots, k$) in the given $m \times n$ rectangle, not counting the square covered by the monomino.

Let

$$m = kq_1 + r_1 \quad (0 \leq r_1 < k),$$

$$n = kq_2 + r_2 \quad (0 \leq r_2 < k).$$

Denote by $P_i (i = 1, 2, \ldots, k)$ the number of unit squares of color i in the entire $m \times n$ rectangle. Simple computations show that

$$P_1 = P_2 = \ldots = P_{r_1} = n\ (q_1 + 1); \tag{2a}$$

$$P_{r_1+1} = P_{r_1+2} = P_k = nq_1. \tag{2b}$$

Since the numbers $S_1, S_2, \ldots, S_k$ are all but one equal to the numbers $P_1, P_2, \ldots, P_k$, and one S_i is exactly one less than the corresponding P_i (we only covered one unit square with the monomino!), we have in view of (1) and (2) exactly two options:

Option 1: $r_1 - 1$ and the monomino is on color 1.

The monomino (let its coordinates be (x, y)) is on color 1; therefore,

$$y \equiv 1 \pmod{k}.$$

Also,

$$m \equiv r_1 \equiv 1 \pmod{k}$$

and

$$P_1 - 1 \equiv P_2 \equiv P_3 \equiv \ldots \equiv P_k \pmod{k},$$

i.e.,

$$n(q_1 + 1) - 1 \equiv nq_1 \pmod{k}$$

or

$$n \equiv 1 \pmod{k}.$$

Thus,

$$m \equiv n \equiv y \equiv 1 \pmod{k}.$$

Option 2: $r_1 = k - 1$ and the monomino is on color k.

The monomino (let its coordinates be (x, y)) is on color k; therefore,

$$y \equiv 0 \pmod{k}.$$

In addition,

$$m \equiv r_1 \equiv -1 \pmod{k}$$

and

$$P_1 \equiv P_2 \equiv \ldots \equiv P_{k-1} \equiv P_k - 1 \pmod{k},$$

i.e.,

$$nq_1 - 1 \equiv n\,(q_1 + 1) \pmod{k},$$

therefore,

$$n \equiv -1 \pmod{k}.$$

Thus, in this case

$$m \equiv n \equiv -1 \pmod{k}$$

and

$$y \equiv 0 \pmod{k}.$$

Now let us rotate the $m \times n$ rectangle 90°, i.e., consider the $n \times m$ rectangle; and apply to it the same reasoning. Our final conclusion is that the following *must* take place

$$\begin{cases} m \equiv n \equiv 1 \pmod{k} \\ x \equiv y \equiv 1 \pmod{k} \end{cases} \quad \text{or} \quad \begin{cases} m \equiv n \equiv -1 \pmod{k} \\ x \equiv y \equiv 0 \pmod{k}. \end{cases}$$

Problem 4.2 is half-solved — we still have to conquer the following two exercises.

Exercise 4.4. Prove that any $m \times n$ rectangle can be tiled with one monomino and linear k-ominoes if $m \equiv n \equiv 1 \pmod{k}$ and the coordinates (x, y) of the monomino satisfy the condition

$$x \equiv y \equiv 1 \pmod{k}.$$

Exercise 4.5. Prove that any $m \times n$ rectangle can be tiled by one monomino and linear k-ominoes if $m \equiv n \equiv -1 \pmod{k}$ and the coordinates (x, y) of the monomino satisfy the condition

$$x \equiv y \equiv 0 \pmod{k}.$$

Martin Gardner, who authored some two dozen magnificent books on popular mathematics, was one of the first readers of this manuscript. I am grateful to him for contributing the following new problem to this book:

Exercise 4.6. (M. Gardner) Prove that the figure L above ("the giant L-tromino," see Figure 4.5) cannot be tiled with linear trominoes.

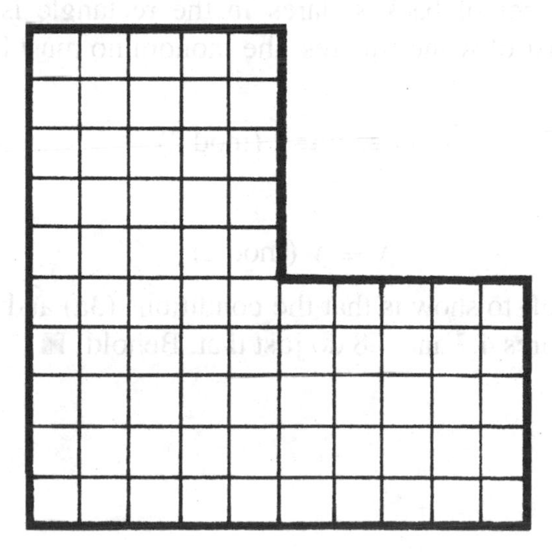

Fig. 4.5 Martin Garner's problem

Solutions to Exercises

4.1. Figure 4.6 solves the problem - behold!

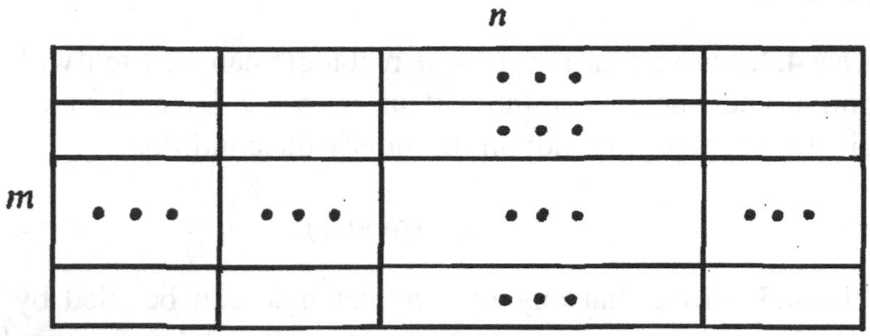

n

m

Fig. 4.6

4.2. The area covered by linear k-ominoes is $mn-1$; it must be divisible by the area of one k-omino, i.e., by k.

4.3. Assume that the $m \times n$ board is tiled with one monomino located at (x, y) and dominoes. Then $mn-1$ must be even; therefore, both m and n must be odd. Color the rectangle in a chessboard fashion using black and white (say, (1,1) is black), and observe that every domino covers exactly one square of each of the colors. Since the number of back squares in the rectangle is one greater than the number of white squares, the monomino must be on a black square, i.e.,

$$m \equiv n \equiv 1 \pmod 2 \tag{3a}$$

and

$$x \equiv y \pmod 2. \tag{3b}$$

All that is left to show is that the conditions (3a) and (3b) are also sufficient. Figures 4.7 and 4.8 do just that. Behold! ■

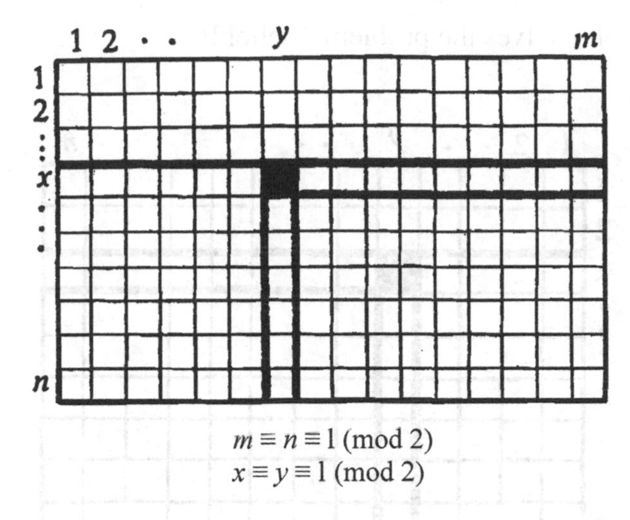

$m \equiv n \equiv 1 \pmod 2$
$x \equiv y \equiv 1 \pmod 2$

Fig. 4.7

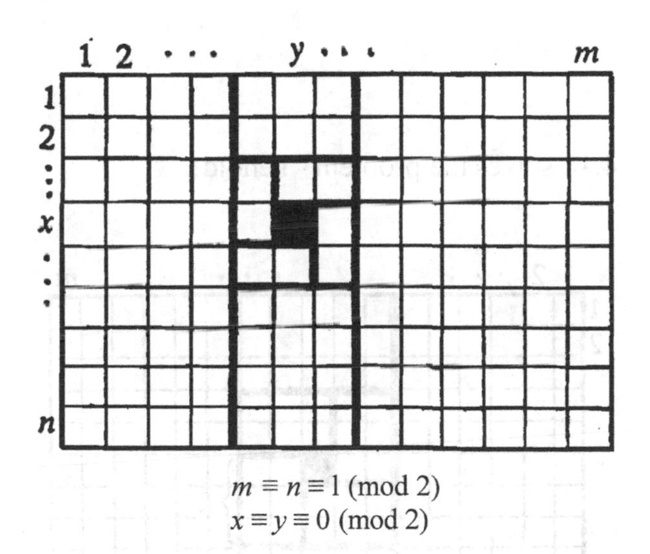

$m \equiv n \equiv 1 \pmod 2$
$x \equiv y \equiv 0 \pmod 2$

Fig. 4.8

4.4. Figure 4.9 solves the problem - behold!

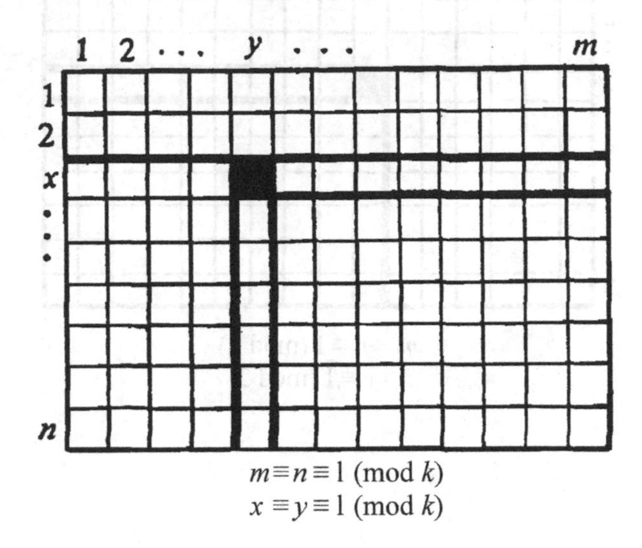

Fig. 4.9

4.5. Figure 4.10 solves the problem - behold!

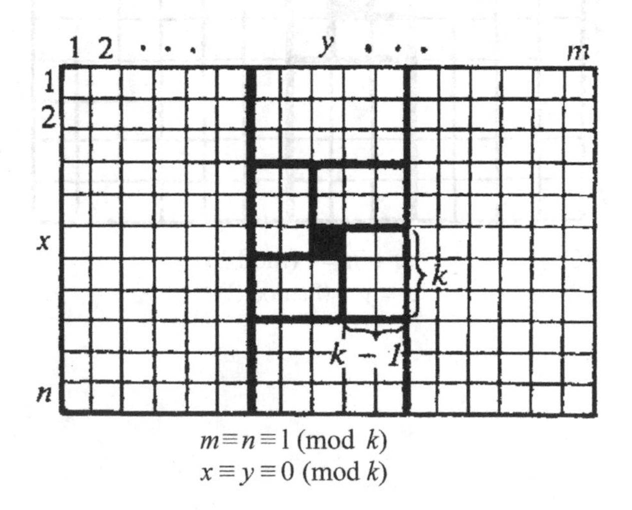

Fig. 4.10

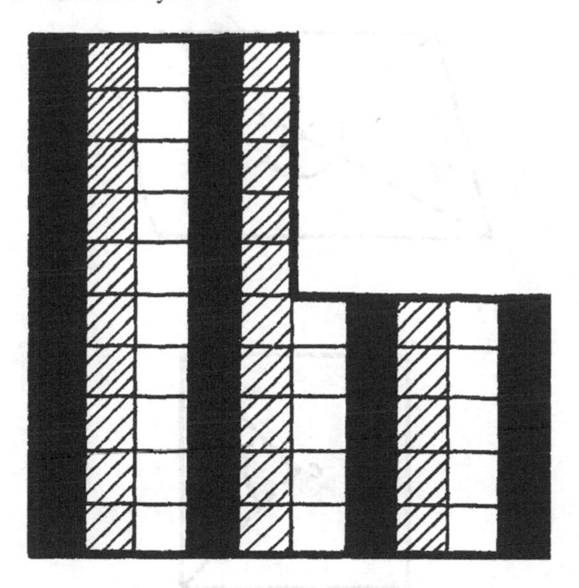

Fig. 4.11

4.6. Let us color the figure L in three colors using the column cyclic coloring (Figure 4.11). If a linear tromino is placed on L horizontally, it covers exactly one square of each color; if a linear tromino is placed on L vertically, it covers three squares of the same color.

Assume that the figure L is tiled by linear trominoes. Then

$$S_1 \equiv S_2 \equiv S_3 \pmod 3, \tag{4}$$

where S_1, S_2, S_3 are the numbers of black, grey, and white unit squares respectively.

On the other hand, $S_1 = 30$; $S_2 = 25$; $S_3 = 20$ which contradicts (4) above. ■

5 Polyominoes and Rotational Symmetries

A figure M is said to possess *n-fold rotational symmetry* if there exists a point c (the *center* of the symmetry or, more precisely, the *n-fold rotational center*), such that M is mapped onto itself under the rotation by $\frac{360°}{n}$ about c. For example, an arbitrary parallelogram possesses 2-fold rotational symmetry, i.e., it is *centrally symmetric* (Figure 5.1).

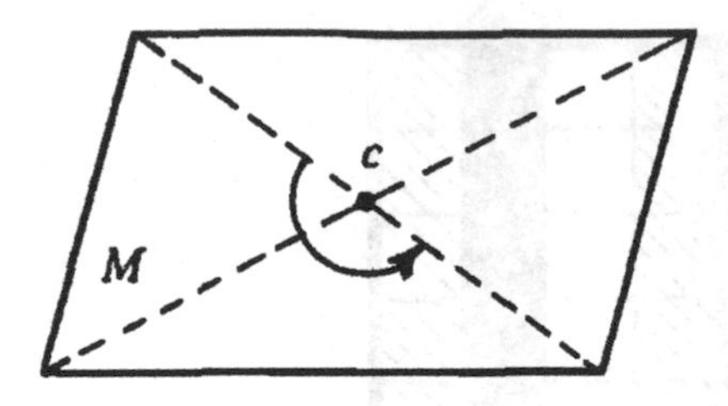

Fig. 5.1

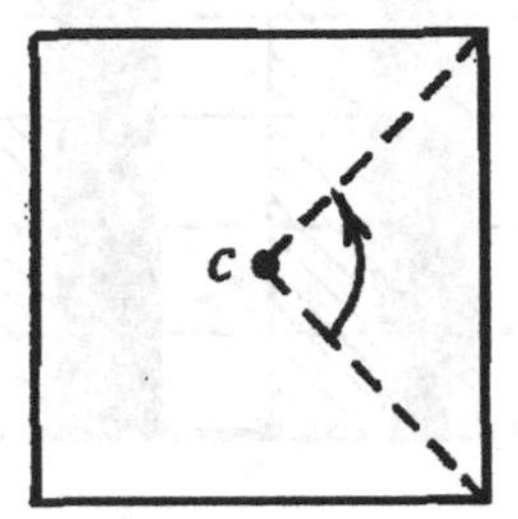

Fig. 5.2

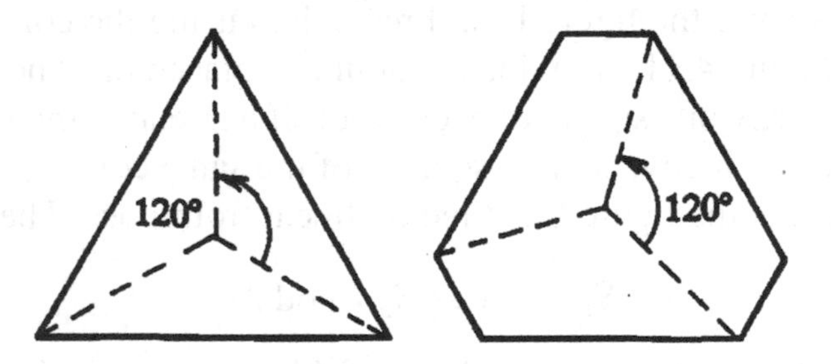

Fig. 5.3

Each square is a figure with 4-fold rotational symmetry (Figure 5.2). The polygons in Figure 5.3 possess 3-fold rotational symmetry. The regular n-gon is a figure with n-fold rotational symmetry.

Now let us look at a rectangle tiled by polyominoes. We say that this *tiling T has* ***n-fold rotational symmetry*** *if there is a point c such that T is mapped onto itself under the rotation by* $\frac{360^\circ}{n}$ *about c.*

In Figures 2.2 and 3.4, the tilings are centrally symmetric, i.e., they possess 2-fold rotational symmetry. In Figure 3.19 the tiling of the 4×4 square by T-tetrominoes has 4-fold rotational symmetry.

Certainly no rectangle M possesses n-fold rotational symmetry except in the case $n = 2$ (and $n = 4$; where M is a square). So, the problem arises: *in what cases does there exist a tiling of a rectangle (or a square) by polyominoes with 2-fold (or 4-fold) rotational symmetry?*

Exercise 5.1. Prove that if both numbers m, and n are odd, then there exists no centrally symmetric tiling of the $m \times n$ rectangle by L-trominoes.

Exercise 5.2. Prove that if m is even and n is divisible by 3, then the $m \times n$ rectangle possesses a centrally symmetric tiling by L-trominoes.

Exercise 5.3. Prove that if n is divisible by 6, then the $5 \times n$ rectangle has a centrally symmetric tiling by L-trominoes.

Exercise 5.4. Prove that if n is divisible by 6, then (for $m \geq 2$) the $m \times n$ rectangle possesses a centrally symmetric tiling by L-trominoes.

Combining the results of Exercises 5.1–5.4 we obtain the following assertion:

Theorem 5.1. *An $m \times n$ rectangle possesses a centrally symmetric tiling by L-trominoes if and only if mn is divisible by 6.*

In the next exercise we consider tilings of L-trominoes with 4-fold rotational symmetry. Of course, such tilings exist only for squares.

Exercise 5.5. Prove that there exists a centrally symmetric tiling of the 6×6 square by L-trominoes.

Exercise 5.6. Prove that an $n \times n$ square possesses a tiling of L-trominoes with 4-fold rotational symmetry if and only if n is divisible by 6.

The above propositions give a complete solution to the problem of tiling of rectangles by L-trominoes with rotational symmetry. Now we look into tilings by L- and T-tetrominoes.

Exercise 5.7. Prove that if m is even and n is divisible by 4, then there exists a centrally symmetric tiling of the $m \times n$ rectangle by L-tetrominoes.

Exercise 5.8. Prove that there is no centrally symmetric tiling of the 3×8 rectangle using L-tetrominoes, but there exists such a tiling for the 3×16 rectangle.

Exercise 5.9. Prove that an $n \times n$ square possesses a tiling by L-tetrominoes with 4-fold rotational symmetry if and only if n is divisible by 4.

Exercise 5.10. Prove that an $n \times n$ square possesses a tiling by T-tetrominoes with 4-fold rotational symmetry if and only if n is divisible by 4.

The authors do not have a proof of the following conjecture:

Conjecture. There is no centrally symmetric tiling of the $m \times 8$ rectangle using L-tetrominoes for any odd m.

Finally, we pose some interesting open problems.

Problem 5.1. Let M be a polyomino and P be a rectangle tiled with copies of M. We say that P is *minimal* if P cannot be decomposed into two or more rectangles, each of which can be tiled by M. For example, the 2×3 and 5×9 rectangles are minimal for the L-tromino (and there are no other minimal rectangles). The question is: *is there, for each given number p, a polyomino M such that there exist more than p minimal rectangles for this polyomino?*

Problem 5.2. A polyomino M consisting of k^2 unit squares has the $2k \times 2k$ square as its minimal rectangle. Is it true that any tiling of the $2k \times 2k$ square by copies of M possesses 4-fold rotational symmetry?

Two L-tetrominoes are said to have the same orientation if there is a rotation that maps one of the L-tetrominoes onto the other.

The existence of two orientations of L-tetrominoes can be illustrated by the following exercise.

Exercise 5.11. One side of the 4×6 rectangular piece R of paper is painted red and the other is left white. Show that R can be cut into L-tetrominoes in such a way that one cannot put them together to form a 3×8 rectangle with one side completely red.

And finally an open problem:

Problem 5.3. Is it true that an $m \times n$ rectangle is tileable by L-tetrominoes of the same orientation if and only if one of the numbers m, n is divisible by 4 and the other is even?[1]

---

[1] See the new to this edition Chapter 5 with a solution of this problem.

Solutions to Exercises

5.1. If both m and n are odd, then there is a *central unit square* in the $m \times n$ rectangle. The center of this unit square coincides with the center of the whole rectangle (Figure 5.4).

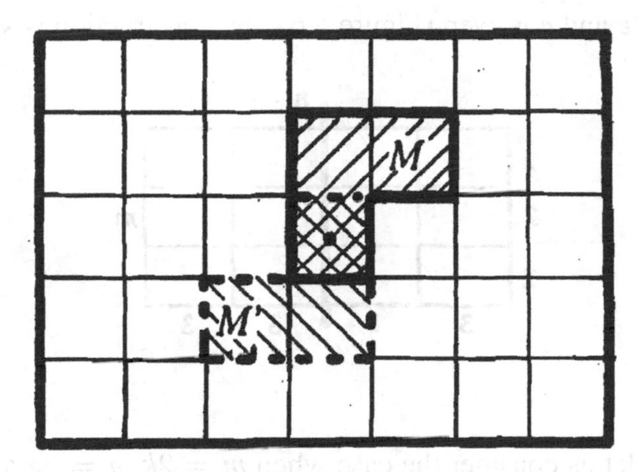

Fig. 5.4

Let M be the L-tromino that contains the central unit square. Then the symmetric L-tromino M' overlaps M. Therefore, there is no centrally symmetric tiling using L-trominoes in this case. ■

5.2. Let us decompose the $m \times n$ rectangle into 2×3 blocks, as shown in Figures 5.5, 5.6, and 5.7.

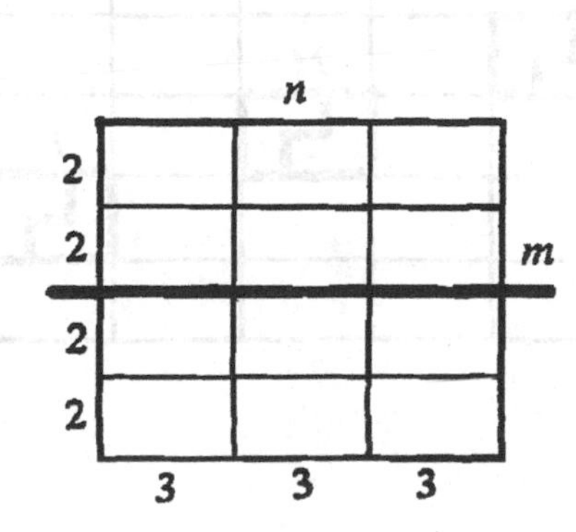

Fig. 5.5

If $m = 2k$ and k is even (as in Figure 5.5), then we can divide the rectangle into two $k \times n$ rectangles, each of which is decomposable into 2×3 blocks. So, constructing a tiling of the upper rectangle with L-trominoes (see Figure 2.2) and taking the symmetric tiling of the lower rectangle (with respect to the center of the whole rectangle), we obtain a centrally symmetric tiling of the whole $m \times n$ rectangle.

If $n = 3e$ and e is even (Figure 5.6), the construction is similar.

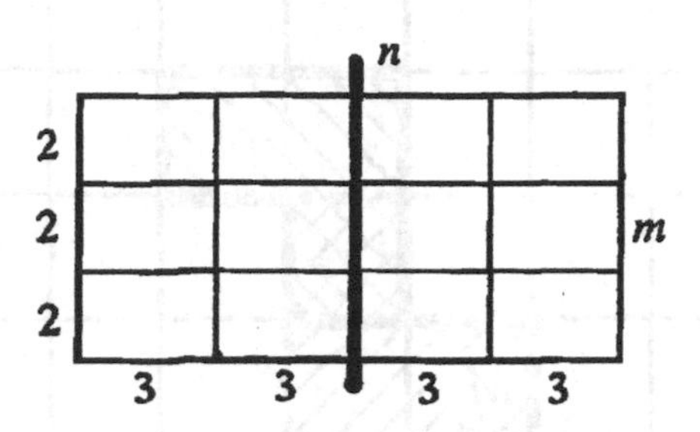

Fig. 5.6

Finally, let us consider the case when $m = 2k, n = 3e$ and both k and e are odd. Then there exists the *central* 2×3 block M in the $m\times n$ rectangle (Figure 5.7), and for any other 2×3 block P, there exists the symmetric block P'. So, choosing a tiling of the central 2×3 block M by L-trominoes (it is centrally symmetric, see Figure 2.2) and choosing symmetric tilings for each pair of symmetric 2×3 blocks P and P', we obtain a centrally symmetric tiling of the whole rectangle. ■

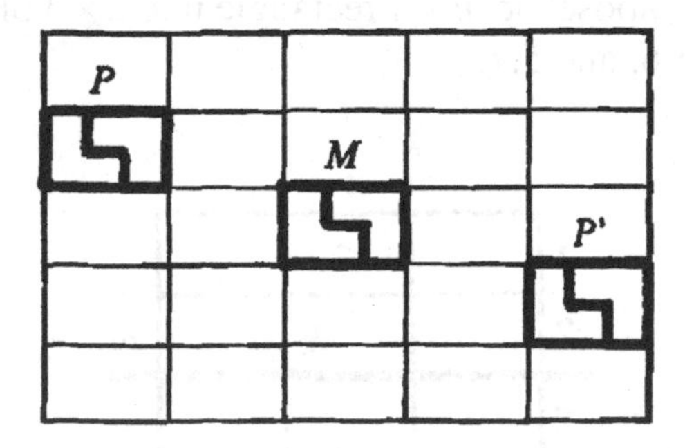

Fig. 5.7

5.3. A centrally symmetric tiling for the 5×6 rectangle is shown in Figure 5.8.

Fig. 5.8

Further, for the $5 \times 6k$ rectangle, we must consider separately the cases when k is even or odd. They are trivial, however (Figure 5.9). ■

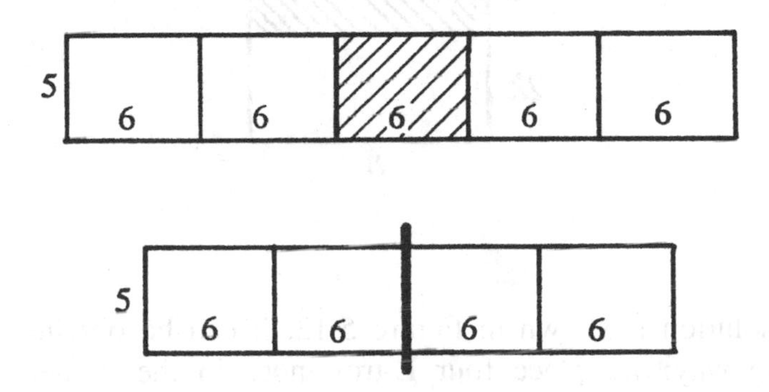

Fig. 5.9

5.4. If m is even, then the required assertion follows directly from the result of Exercise 5.2. Now let m be odd. Then either $m = 4k + 3$, or $m = 4k + 5$, for $k \geq 0$. Therefore, we can decompose the $m \times n$ rectangle into either one central $3 \times n$ block and two $2k \times n$ blocks (Figure 5.10) or into one central $5 \times n$ block and two $2k \times n$ blocks (Figure 5.11). In view of Exercises 5.2 and 5.3, and using the divisibility of n by 6, this completes the solution. ■

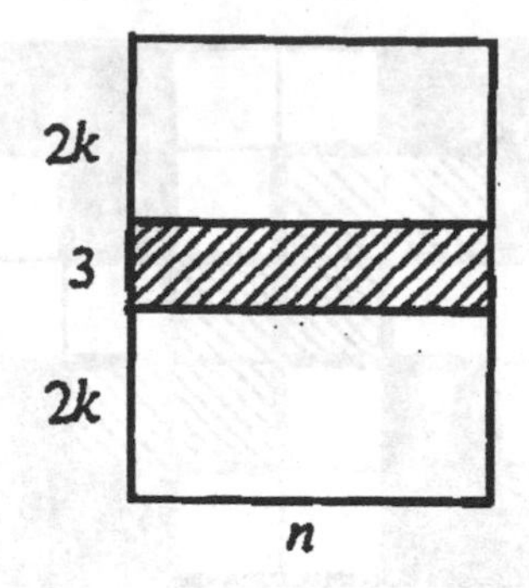

Fig. 5.10

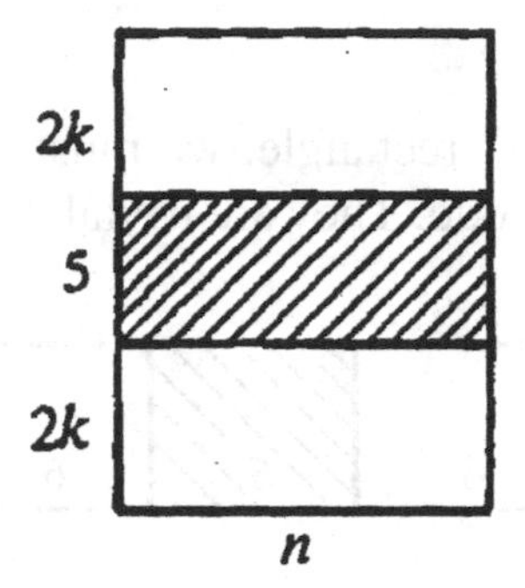

Fig. 5.11

5.5. A solution is shown in Figure 5.12. It can be obtained in the following way: we place four L-trominoes in the corners of the 6×6 square and tile of the central 4×4 square with T-tetrominoes

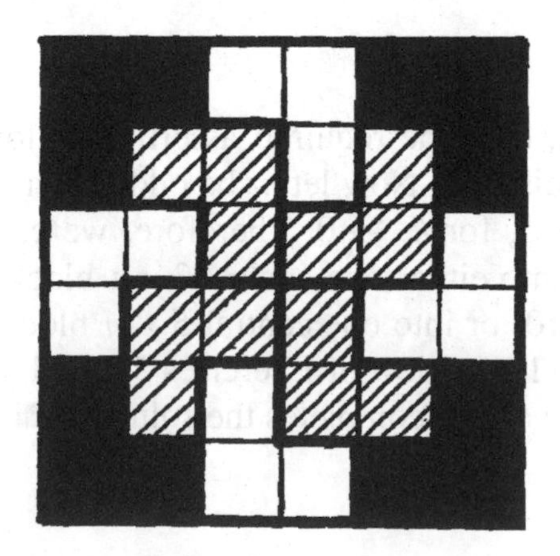

Fig. 5.12

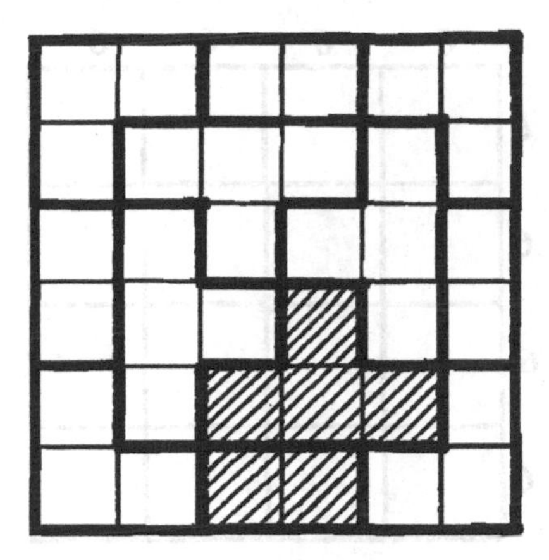

Fig. 5.13

(Figure 5.13). This tiling (by four dominoes, four L-trominoes, and four T-tetrominoes) possesses 4-fold rotational symmetry. In order to get the tiling of Figure 5.12, it remains now to "correct" the tiling in Figure 5.13 by replacing the union of one domino and one tetromino (shaded in Figure 5.13) by two L-trominoes. ■

5.6. By Theorem 5.1, if the $n \times n$ square possesses a tiling by L-trominoes with 4-fold rotational symmetry (and consequently, with 2-fold rotational symmetry), then n^2 is divisible by 6. But then n is divisible by 6 as well.

Conversely, if n is divisible by 6, that is, $n = 6k$, then the $n \times n$ square can be decomposed into 6×6 blocks. Now (separately for even and odd k, see Figures 5.14 and 5.15), we can construct a tiling of the $n \times n$ square by L-trominoes with 4-fold rotational symmetry (see Figure 5.12). ■

5.7. Let $m = 2k$ and $n = 4e$. If k is even, then we can decompose the $m \times n$ rectangle into 2×4 blocks in a way that is centrally symmetric (Figure 5.16). Figure 3.4 shows how to complete the required tiling. Similarly, if e is even, we can construct the required tiling (Figure 5.17). ■

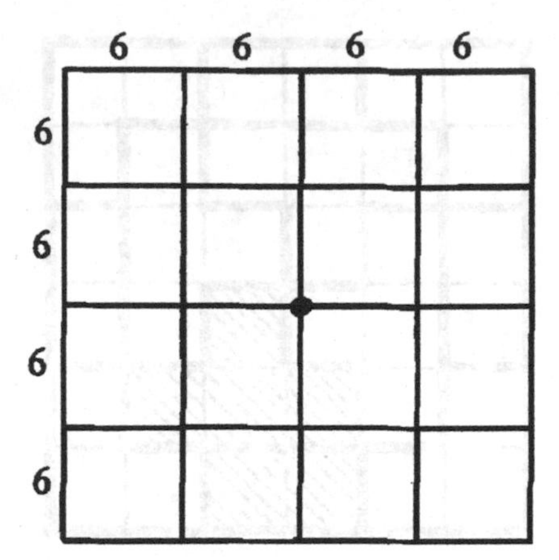

Fig. 5.14

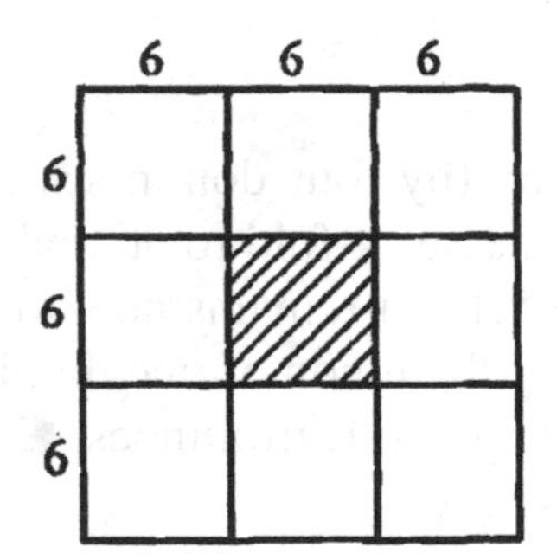

Fig. 5.15

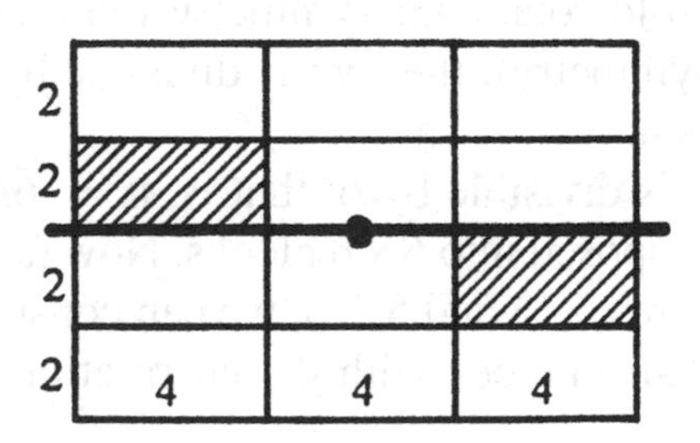

Fig. 5.16

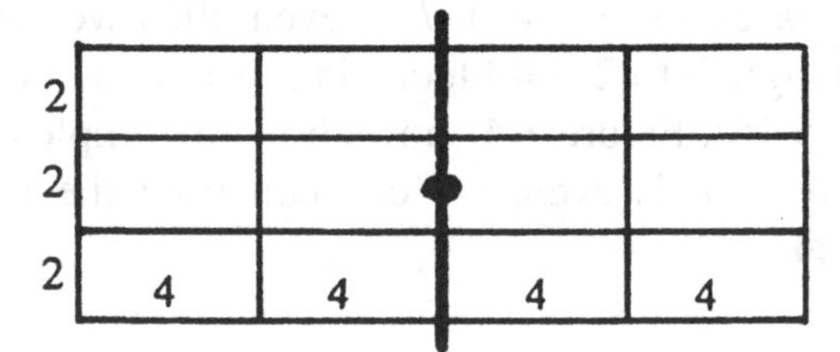

Fig. 5.17

Finally, if both k and e are odd, then in the decomposition into 2×4 blocks there is the *central* block (Figure 5.18), and for each noncentral 2×4 block P there is the symmetric block P'. This allows us to construct a centrally symmetric tiling of the rectangle using L-tetrominoes (similarly to the solution of Exercise 5.2). ■

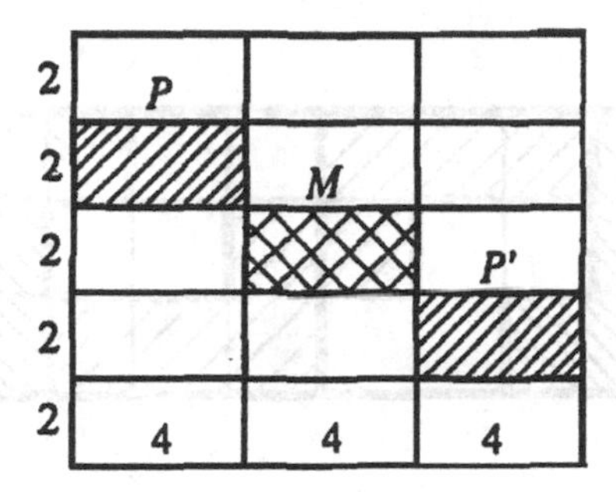

Fig. 5.18

5.8. There are three ways to cover the upper left corner of the 3×8 rectangle (in Figure 5.19a, each time we used a T-tetromino we drew a symmetric one).

In Figure 5.19(a) we have two possible continuations (Figures 5.20a and 5.20b), but neither of them allows for the completion of tiling.

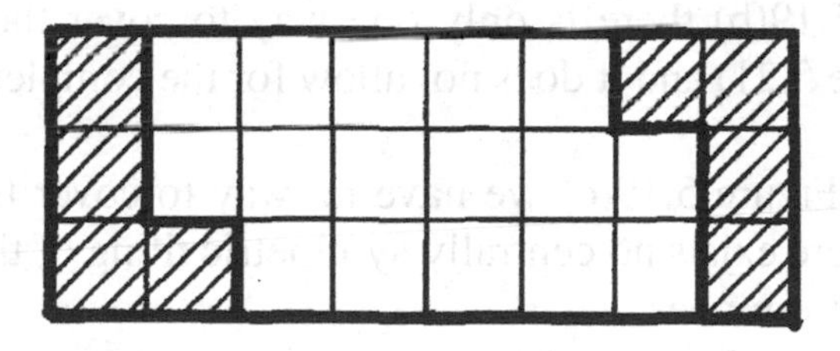

Fig. 5.19(a)

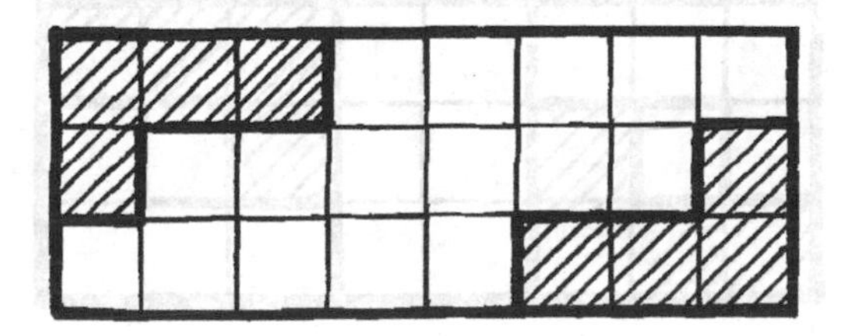

Fig. 5.19(b)

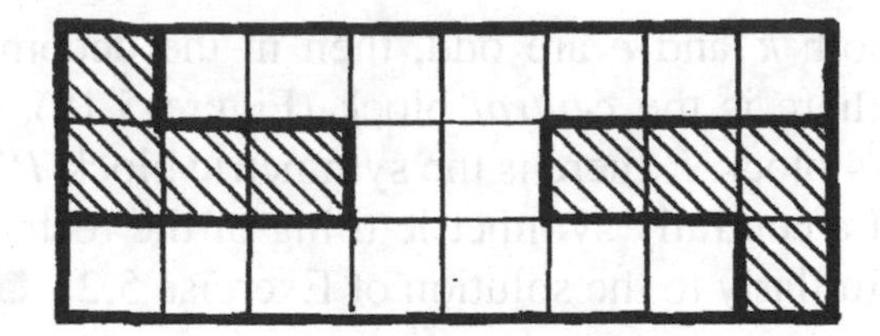

Fig. 5.19(c)

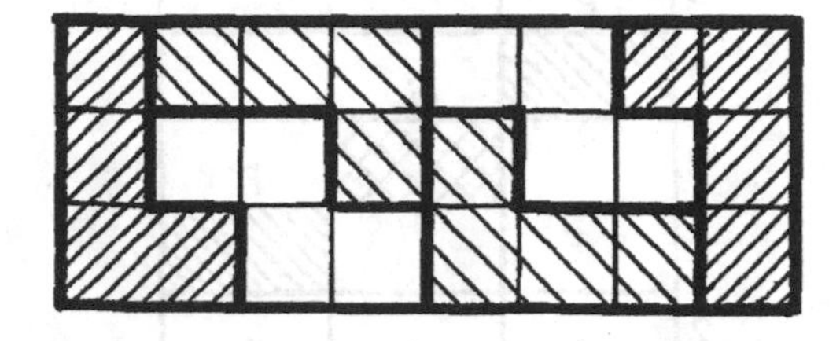

Fig. 5.20a

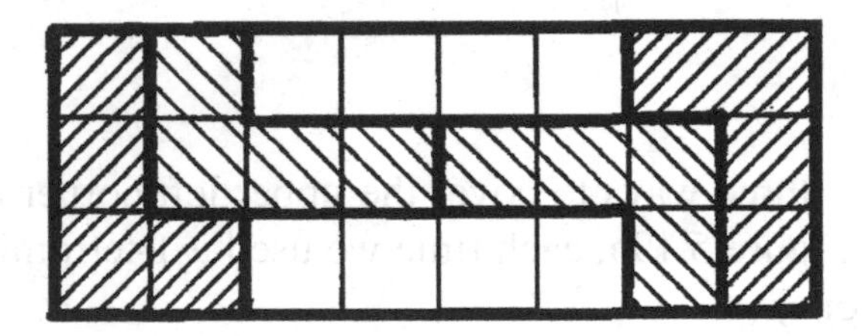

Fig. 5.20b

In Figure 5.19(b) there is only one way to cover the bottom left corner (Figure 5.21) and it does not allow for the completion of tiling, either.

Finally, in Figure 5.19(c) we have no way to cover the lower left corner. So, there exists no centrally symmetric tiling of the 3×8 rectangle by L-tetrominoes.

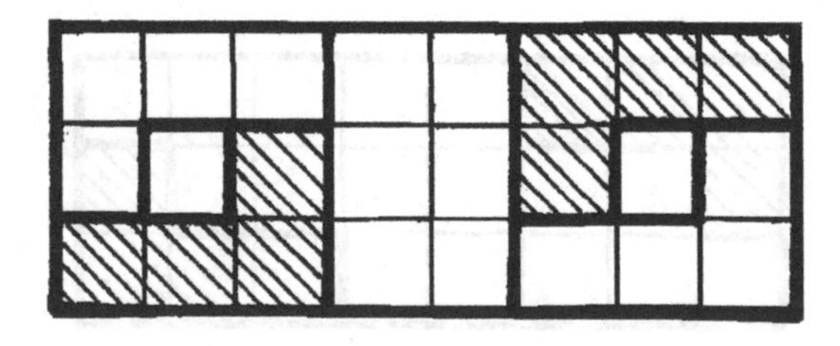

Fig. 5.21

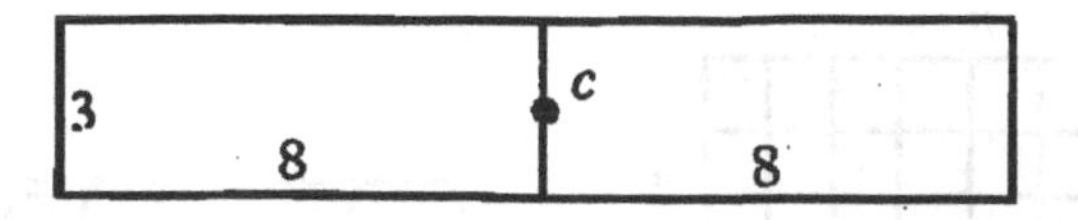

Fig. 5.22

As for a 3×16 rectangle, we can take a tiling of the 3×8 rectangle in the left half of Figure 5.22 and then construct the tiling in the right part symmetric to the former one with respect to point c. Thus we obtain a centrally symmetric tiling of the 3×16 rectangle using L-tetrominoes. ■

5.9. Exercise 3.8 of Section 3 implies that if an $n \times n$ square possesses the required tiling, then n^2 is divisible by 8. Consequently, n is divisible by 4.

Conversely, if n is divisible by 4, then the $n \times n$ square can be decomposed into 4×4 blocks, and consequently, this square possesses a centrally symmetric tiling by L-tetrominoes. ■

5.10. The solution is similar to the solution to Exercise 5.9. ■

5.11. Cut R into L-tetrominoes of the same orientation. Then show that the 3×8 rectangle cannot be tiled using L-tetrominoes of the same orientation. ■

6 Tiling on Other Surfaces

In the preceding sections we tiled a plane $m \times n$ rectangle R. We can glue together two opposite sides of R and thus form a *cylinder* $C\,(m \times n)$. In fact, we can go further and glue together the two bases of the cylinder and thus form a donut (or bagel, depending upon your taste). Mathematicians call this a *torus* (Figure 6.1), so we denote it by $T(m \times n)$.

Of course, if the $m \times n$ rectangle is tileable by, say, tiles of shape S, so is the cylinder $C\,(m \times n)$. And if the cylinder $C(m \times n)$ is tileable, so is the torus $T(m \times n)$. But the converse is not always true. Sometimes cylinders are "better tileable" than the original rectangle, and in turn, the torus can be "better tileable" than the cylinder. Please also note that the $m \times n$ rectangle gives birth to two cylinders,

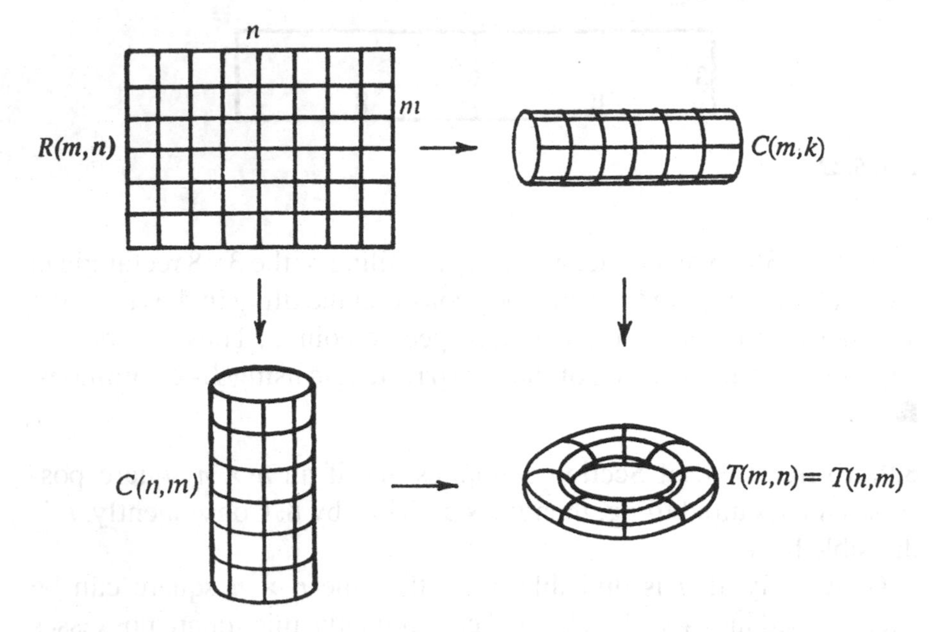

Fig. 6.1 Making two cylinders and one torus out of a rectangle

$C(m \times n)$ and $C(n \times m)$, depending upon which pair of opposite sides is glued together. We, however, get only one torus $T(m \times n)$. Indeed, in order to obtain the torus $T(m, n)$ from the original $m \times n$ rectangle directly, we glue together both pairs of opposite sides of the rectangle simultaneously.

Now you are ready to explore on your own!

Exercise 6.1. Find all n such that the torus $T(n \times n)$ can be tiled with L-trominoes.

Exercise 6.2. Can it so happen that the cylinder $C(m \times n)$ is tileable by T-tetrominoes, but the cylinder $C(n \times m)$ is not?

Exercise 6.3. Find all m and n such that the cylinder $C(m \times n)$ can be tiled with linear k-ominoes. Can it so happen that exactly one of the cylinders $C(m \times n)$, $C(n \times m)$ can be tiled with linear k-ominoes?

Let us go back to our original $m \times n$ rectangle. If we first twist the rectangle and then glue its opposite sides, we get what is known as the *Möbius band* $M(m \times n)$ (Figure 6.2). We imagine this surface as a mathematical abstraction without thickness. It is a very unusual surface. If you travel from any point P along the Möbius band you

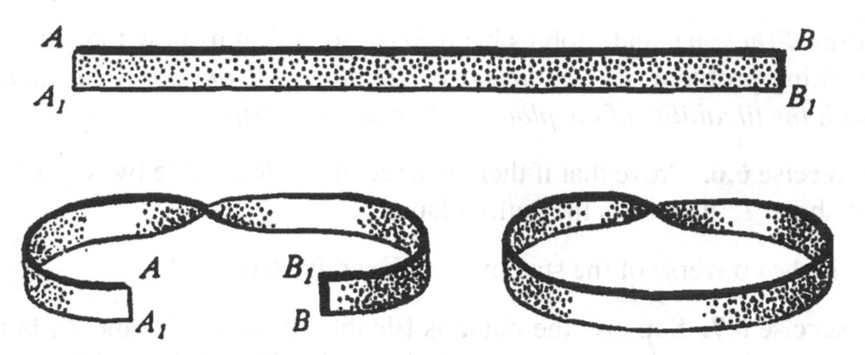

Fig. 6.2 Making a Möbius band

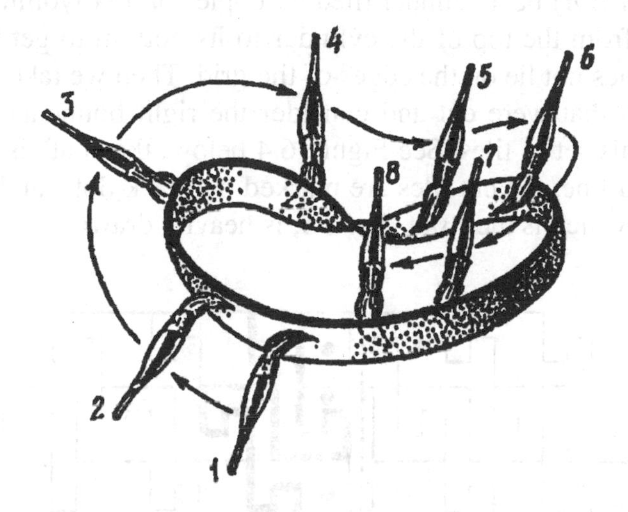

Fig. 6.3

will get back to P but you will be situated on the other side of the Möbius band (Figure 6.3).

Exercise 6.4. We know that the 4×9 rectangle cannot be tiled by L-tetrominoes. Prove that the Möbius band $M(4 \times 9)$ can be tiled by L-tetrominoes.

Exercise 6.5. Let P be a polyomino. It is known that exactly one of the two Möbius bands $M(m \times n)$, $M(n \times m)$ can be tiled by copies of P. What can you conclude about the tileability of the plane $m \times n$ rectangle?

So far in this section we have compared the tileability of an $m \times n$ rectangle with the tileability of the corresponding cylinder $C(m \times n)$, torus $T(m \times n)$, and Möbius band $M(m \times n)$. Let us now look at the exciting problem of *comparing the tileability of the above surfaces with the tileability of the plane and an infinite strip.*

Exercise 6.6. Prove that if there is a rectangle R tileable by, say, tiles of shape T, then so is the entire plane.

Is the converse of the statement of Exercise 6.6 true?

Exercise 6.7. Suppose the plane is tileable by tiles of shape T. Is it true that there exists a rectangle R tileable by tiles of shape T?

Let $C(m \times n)$ be a cylinder tiled by copies of a polyominal tile T. We "cut" from the top of the cylinder to its bottom to get a vertical line that does not lie on the edges of the grid. Then we take the union of the tiles that were cut and consider the right boundary S of the union of this set of tiles. See Figure 6.4 below: the "cut" is indicated by a dotted line, the cut tiles are marked by black dots, and the right boundary, which is the *"step-line"* S, is heavily drawn.

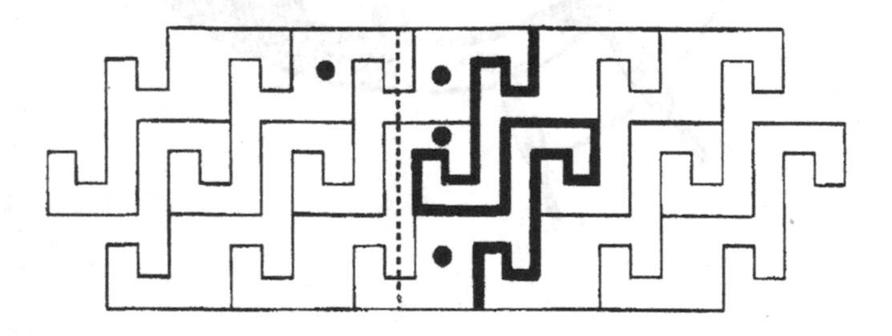

Fig. 6.4

Now we cut the cylinder along S, and open up and flatten this surface. We get a plane figure F with parallel top and bottom boundaries and "parallel" left and right "step-line" boundaries. Now it is clear how to formalize these notions:

The *step-line* is the one that can be traced continuously along the lines of a checkered grid without self-intersection.

We will call two step-lines *parallel* if one of them is an image of the other under translation.

The opposite sides of the figure F are parallel. Let t be a translation that maps the left boundary of F onto its right boundary. Then by repeatedly applying the translation t and the *inverse* translation t^{-1}

(i.e., a move in the opposite direction through the same distance) to F and its consequent images, we tile an infinite strip S with copies of F (Figure 6.5).

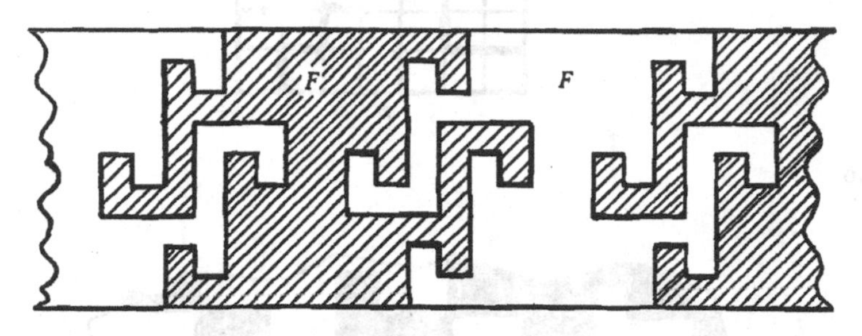

Fig. 6.5

Since the figure F itself is tiled with tiles of shape T, we get a tiling of the infinite strip S with tiles T. This type of tiling of an infinite strip using tiles T is called *periodic:* it is obtained by repeating a translation t (and t^{-1}) of a tiling of a bounded figure F.

We proved that *every tiling of α cylinder leads to a periodic tiling of an infinite strip.* Now you are ready for two very exciting questions that were contributed by the famous geometer, Branko Grünbaum, a coauthor of the great book, *Tiling and Patterns* ([GS]).

Exercise 6.8. (B. Grünbaum) Is there a tiling of an infinite strip by copies of a tile T that is not periodic?

Exercise 6.9. (B. Grünbaum) Prove that every tiling of an infinite strip by copies of a polyominal tile T contains a bounded part that can be used to tile a cylinder. (Hint: read the first paragraph of Section 8).

Let $T(m \times n)$ be a torus tiled by copies of a polyominal tile T. We can picture the torus $T(m \times n)$ as the plane $m \times n$ rectangle R but we must remember that the opposite sides of R are glued together. Now we can cut the torus by two step-lines along the boundaries of some adjacent *tiles*, so that one step-line goes from the top to the bottom edge of R, and the other from the left to the right edge of R (Figure 6.6).

Now we combine the four parts of R from Figure 6.6 together to form a figure F (see Figure 6.7).

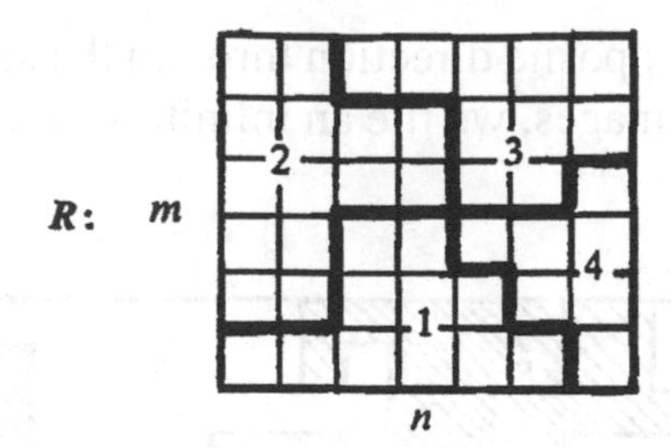

Fig. 6.6

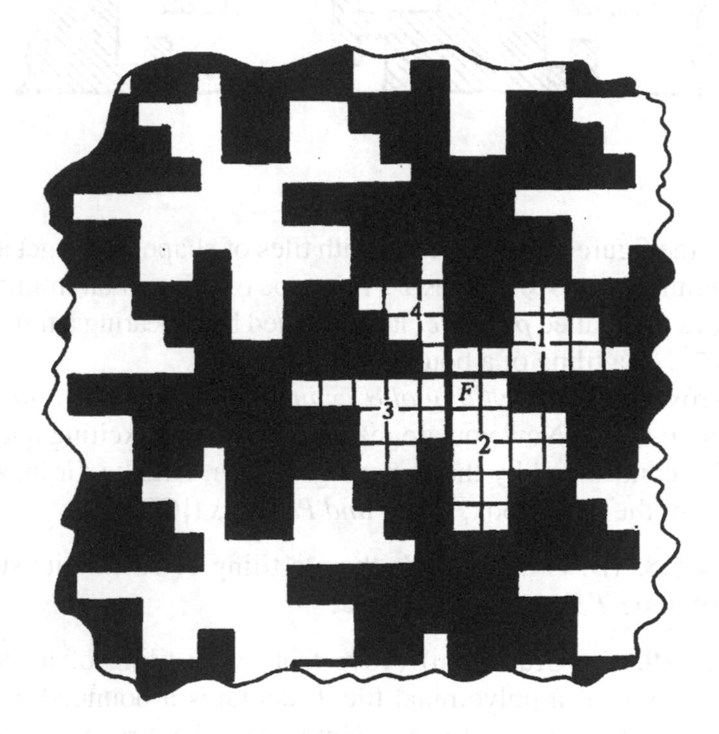

Fig. 6.7

Let t_1 be the translation to the right through n and t_2 the translation upward through m. The figure F has two useful properties:

i) By repeatedly applying the translations t_1 and t_2 and their inverses t_1^{-1} and t_2^{-1} to F and its consequent images, we will tile the entire plane with copies of F.
ii) F can be tiled using copies of tile T (that originally tiled the torus $T(m, n)$).

We leave the pleasure of proving properties i) and ii) for the readers. These two properties give birth to a tiling of the plane with copies of tile T. This type of tiling of the plane is called *periodic:* it is obtained by repeating applications of the translations t_1 and t_2 (and their inverses) to a tiling of a bounded figure F.

We proved here that *every tiling of a torus leads to a periodic tiling of the plane.*

We do not know the answer to the following question contributed by Branko Grünbaum. The only consolation is that apparently nobody does!

Grünbaum's Open Problem 6.1. Is it true that every tiling of the plane by copies of a polyominal tile contains a bounded part that can be used to tile a torus?

We are offering \$50 for the first solution to this problem. (The authors will be the sole judges of what constitutes a solution.)

Tiling a plane rectangle can be expanded in a different way by adding a dimension (or a few!) to our tiling games. In three dimensions we call this *packing* a parallelepiped (Figure 6.8).

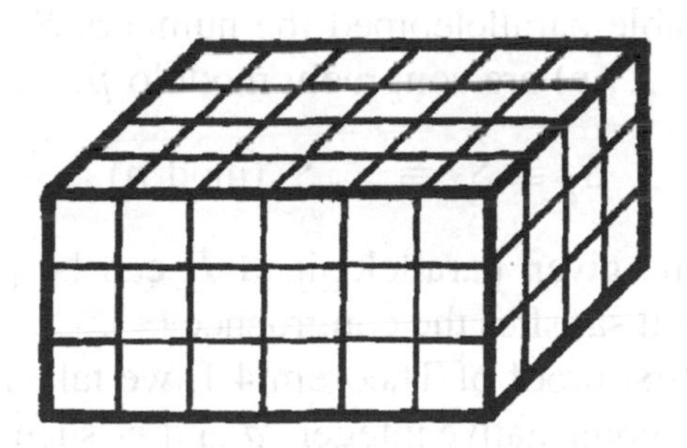

Fig. 6.8

Example 6.1. Find all positive integers m, n, and k such that the parallelepiped $m \times n \times k$ can be packed with linear 3-dimensional p-ominoes (i.e., $1 \times 1 \times p$ bricks, see Figure 6.9).

Solution: Assume that an $m \times n \times k$ parallelepiped P can be packed with linear 3-dimensional p-ominoes, but neither m nor n are divisible by p. We will prove that k is divisible by p.

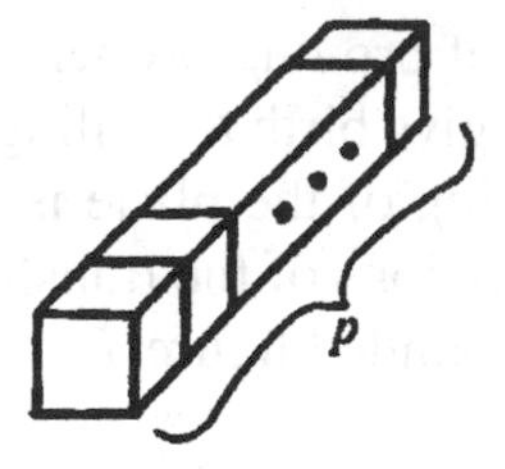

Fig. 6.9

Let us color all mnk unit squares of P in p colors using a coloring we call ***the direct product of diagonal and column colorings*** or the ***$D \times C$ coloring*** for short (see Figures 4.1, 4.2, and 4.3). In other words, we color the lower horizontal $m \times n$ layer of unit cubes diagonally in p colors with cyclic permutation of colors (see Figures 4.2 and 4.1). The parallelepiped can be viewed as a union of mn vertical columns, each "growing" up from its foundation, i.e., from a unit cube of the lowest level. We color all the unit cubes of a column in the same color as their foundation; thus we get mn one-color columns.

You undoubtedly noticed that in a $D \times C$ coloring, a linear p-omino placed horizontally covers one unit cube of each of the p colors, and each vertical linear p-omino covers p cubes of the same color. Therefore, for any tileable parallelepiped the numbers S_i of unit cubes of color $i\,(i = 1, 2, \ldots, p)$ are congruent modulo p:

$$S_1 \equiv S_2 \equiv \ldots S_p (\text{mod } p). \qquad (*)$$

Assume that the given parallelepiped P can be packed by linear p-ominoes. Then it satisfies the congruence $(*)$.

Just as in the first proof of Theorem 4.1, we take advantage of the fact that there are nonnegative integers q and r_1 such that $0 \le r_1 \le \frac{p}{2}$ and

$$m = pq + r_1$$

or

$$m = pq - r_1.$$

Accordingly, we consider two cases:

Case 1: $m = pq + r_1$. We cut the parallelepiped P into two parallelepipeds: $pq \times n \times k$ and $r_1 \times n \times k$. Since the $pq \times n \times k$ parallelepiped can be packed with linear p-ominoes, it satisfies the condition $(*)$. Since the congruence $(*)$ is true for the parallelepiped P, it must also be true for the $r_1 \times n \times k$ parallelepiped P_1.

Case 2: $m = pq - r_1$. Let us attach the $r_1 \times n \times k$ parallelepiped P_1 to the given parallelepiped P to obtain the $pq \times n \times k$ parallelepiped P' and extend the coloring of P to the $D \times C$ coloring of P'. Since P' can be packed by linear p-ominoes, P' satisfies the condition $(*)$. Since both P' and P satisfy condition $(*)$, their difference, the parallelepiped P_1, satisfies the condition $(*)$ as well.

Thus, in both cases we found an $r_1 \times n \times k$ parallelepiped P_1 such that $0 \le r_1 \le \frac{p}{2}$ that satisfies the condition $(*)$. Now we can rotate P_1 about a vertical axis through a 90° angle, i.e., consider the $n \times r_1 \times k$ parallelepiped, and apply to it *all* of the above reasoning. As a result, we get the $r_2 \times r_1 \times k$ parallelepiped P_2 with $0 \le r_1 \le \frac{p}{2}$ that satisfies the condition $(*)$.

On the other hand, the number of one-color unit squares in the $r_2 \times r_1$ rectangle is equal to $r_2 + r_1 - 1$ and

$$r_2 + r_1 - 1 \le \frac{p}{2} + \frac{p}{2} - 1 < p.$$

Therefore, the p-th color is not present in the parallelepiped P_2 at all; i.e.,

$$S_p = 0,$$

while we have exactly one column of the first color, i.e.,

$$S_1 = k.$$

The above two equalities together with the congruence $(*)$ show that $k \equiv 0 \pmod{p}$, i.e., k is divisible by p.

Thus a parallepiped $m \times n \times k$ can be packed by linear 3-dimensional p-ominoes if and only if at least one of the numbers m, n, k is divisible by p. ■

Exercise 6.10. Find all positive integers m, n, and k such that the $m \times n \times k$ parallelepiped can be packed by 3-dimensional L-trominoes (Figure 6.10).

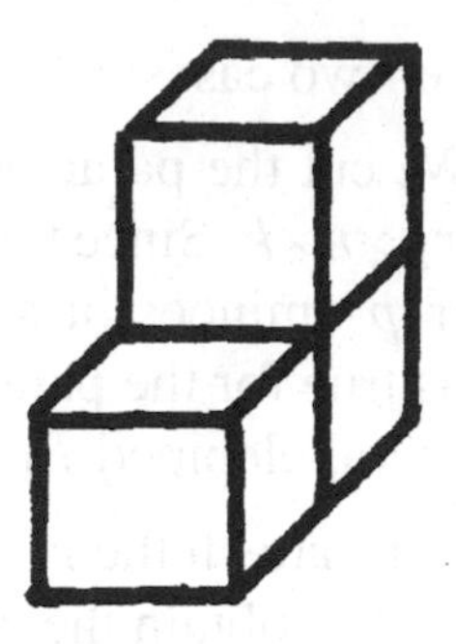

Fig. 6.10

Exercise 6.11. Let k and n be positive integers, $k < n$. Prove that no matter which $kn^2 + 1$ unit cubes of the $n \times n \times n$ cube C are colored red, you can always choose $k + 1$ red cubes such that no two of them lie in the same layer of C. (There are three different types of layers in C that are parallel to various faces of C.)

In fact, Exercise 6.11 allows an n-dimensional generalization that was noticed by A. Soifer and S. Slobodnik in 1973 (see [SS], [S1], [S9]), and appears with a proof and related problems in [S4]. For $n = 10$ this generalization can be nicely reformulated in the language of numbers using the decimal system:

Exercise 6.12. (A. Soifer and S. Slobodnik) Given $r \cdot 10^{n-1} + 1$ distinct n-digit numbers, $0 < r < 9$. Prove that you can choose $r + 1$ numbers out of them such that any two of the $r + 1$ numbers in any decimal location have distinct digits.

Solutions to Exercises

6.1. As we know from Section 2, an $n \times n$ square can be tiled by L-trominoes if and only if n is divisible by but not equal to 3. On the torus, this exception is not necessary: the torus $T(3 \times 3)$ can be tiled (Figure 6.11).

Indeed, when we form the torus from the 3×3 square, the three black squares get together and form an L-tromino! ■

Fig. 6.11 Tiling the 3×3 torus with L-trominoes

6.2. Yes. The cylinder $C(4 \times 2)$ can be tiled by T-tetrominoes (Figure 6.12). The cylinder $C(2 \times 4)$ cannot be tiled.

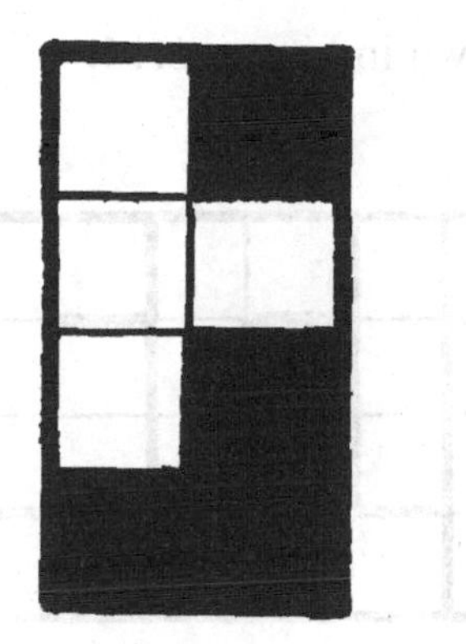

Fig. 6.12

Indeed, when we glue the top and bottom of the 4×2 rectangle together, the black squares come together and form a T-tetromino (Figure 6.12). The cylinder $C(2 \times 4)$ cannot be tiled because two squares in the upper row of the 4×2 rectangle must be covered by two distinct T-tetrominoes T_1 and T_2, which in turn forces T_1 and T_2 to overlap. ■

6.3. The cylinder $C(m \times n)$ can be tiled by linear k-ominoes if and only if at least one of m, n is divisible by k (in other words, this is no different than tiling a plane $m \times n$ rectangle).

To prove this, we color rings of the $C(m \times n)$ cylinder in k colors with cyclic permutation of colors (Figure 6.13), and then proceed exactly as in the second solution to Problem 4.1 in Section 4. ■

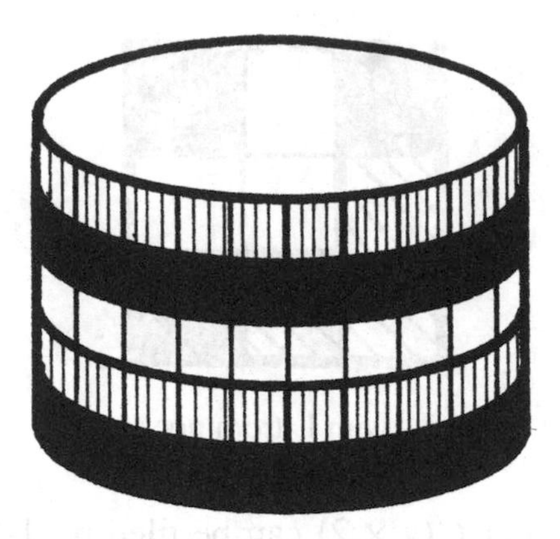

Fig. 6.13

6.4. A solution is shown in Figure 6.14.

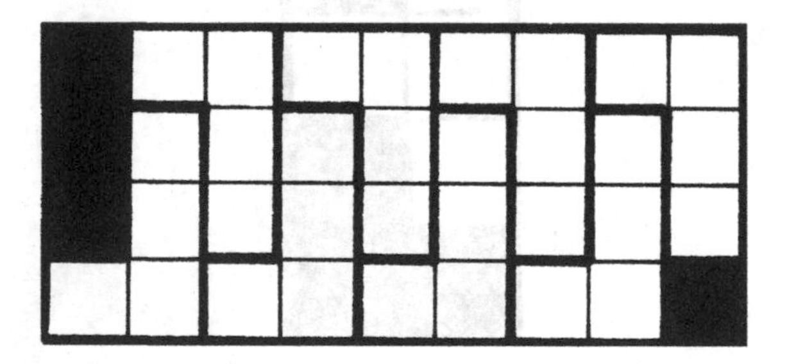

Fig. 6.14

When we glue the Möbius band from this rectangle, the two black regions join together, creating an L-tetromino. ■

6.5. Assume that the $m \times n$ rectangle can be tiled by copies of a polyomino P. Then both Möbius bands $M(m \times n)$ and $M(n \times m)$ will be tileable with copies of P. But we are given that exactly one of them is tileable. Therefore, the $m \times n$ rectangle cannot be tiled with copies of P. ■

6.6. Copies of R easily tile the plane (Figure 6.15). Tiling each copy of R with tiles of shape T will thus generate a tiling of the plane using tiles of shape T. ■

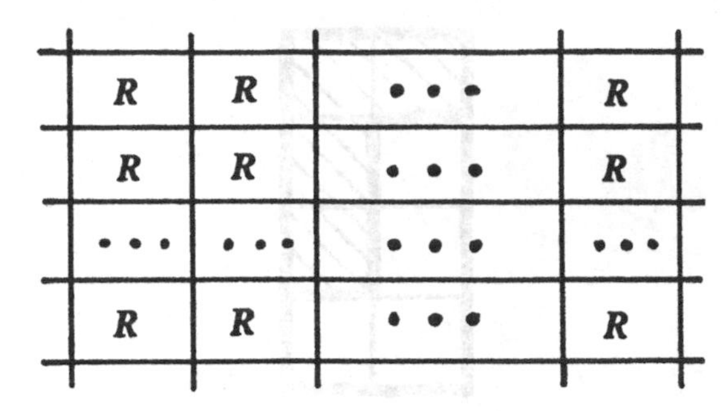

Fig. 6.15

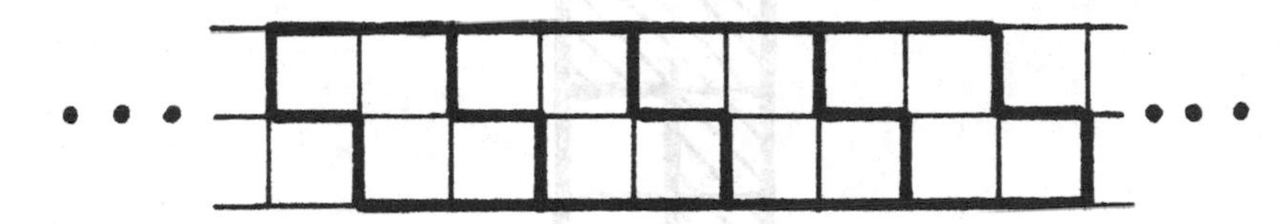

Fig. 6.16

6.7. No! We can tile an infinite strip using Z-tetrominoes (Figure 6.16).

Such strips easily tile the plane. On the other hand, in Exercise 3.2 we proved that there is no rectangle tileable by Z-tetrominoes. ■

6.8. There is! We can split the 4-wide infinite strip into 4×2 rectangles, and put in each of them either "0" or "1," alternating as follows: . . . 10000100010010100010001 00001 . . . (Figure 6.17).

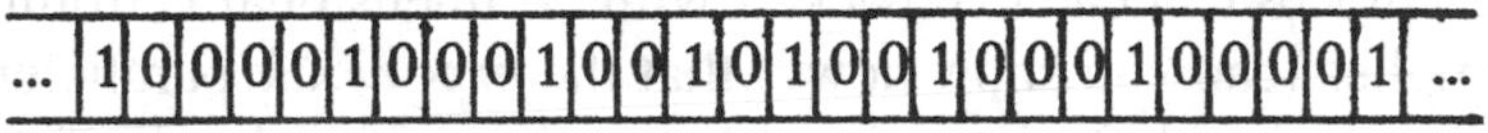

...	1	0	0	0	0	1	0	0	0	1	0	0	1	0	1	0	0	1	0	0	0	1	0	0	0	0	1	...

Fig. 6.17

Now we tile all the 4×2 rectangles with "0" as in Figure 6.18(a). All the 4×2 rectangles with "1" we tile as shown in Figure 6.18(b). As a result we get a nonperiodic tiling of the infinite strip by L-tetromino (can you prove it?). ■

Fig. 6.18(a)

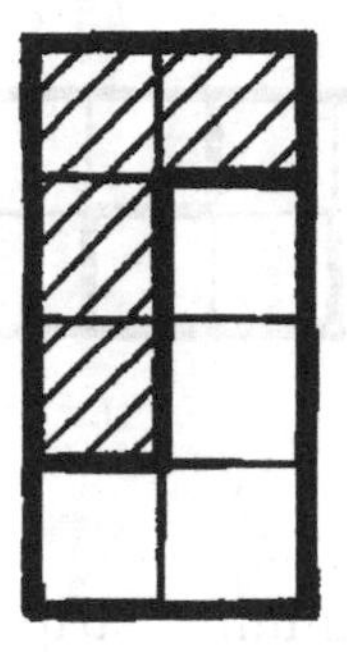

Fig. 6.18(b)

6.9. Please see Exercise 8.5. ■

6.10. Let $m \geq n \geq k$. If $k = 1$, then the solution for m and n is exactly the same as in Theorems 2.1 and 2.2 of Section 2.

Let $k \geq 2$. You can show (do) that the $3\times2\times2$, $3\times3\times2$, and $3\times3\times3$ parallelepipeds can each be packed with 3-dimensional L-trominoes. This implies (can you show how?) that the $m \times n \times k$ parallelepiped can be packed with 3-dimensional L-trominoes if and only if mnk is divisible by 3. ■

6.11. This problem is a 3-dimensional analog to Exercise 7.6 discussed in the next section. ■

Chapter 2
Proofs of Existence

7 The Pigeonhole Principle in Geometry

We start by offering the reader the opportunity to prove the *Pigeonhole Principle* itself. It is also known (in Europe) as the *Dirichlet Principle*, after the famous mathematician Peter Gustav Lejeune Dirichlet (1805–1859).

Exercise 7.1. (Pigeonhole Principle) If $kn + 1$ pigeons (where k and n are positive integers) sit on n pigeonholes, then at least one of the holes has at least $k + 1$ pigeons on it.

Note that the principle guarantees the *existence* of a hole with lots of pigeons on it, but as often happens in mathematics, it gives us no way of finding this hole.

Example 7.1. Prove that among any five points located inside or on the boundary of a unit square there are two points at most $\frac{1}{\sqrt{2}}$ apart.

Solution. As you can see, *pigeons and pigeonholes are not given to us*. We have to invent them if we wish to use the Pigeonhole Principle.

Let us partition the given unit square into four quarter squares (Figure 7.1). These quarter squares will be our pigeonholes, and, of course, the five given points will serve as the pigeons.

Since $5 = 1 \times 4 + 1$, by the Pigeonhole Principle there is a pigeonhole that contains at least two pigeons. In other words, there is a quarter square that contains at least two given points A, B. The distance $|AB|$ is, of course, no greater than the diagonal $\frac{1}{\sqrt{2}}$ of the quarter square. ■

A. Soifer, *Geometric Etudes in Combinatorial Mathematics*,
DOI 10.1007/978-0-387-75470-3_2,

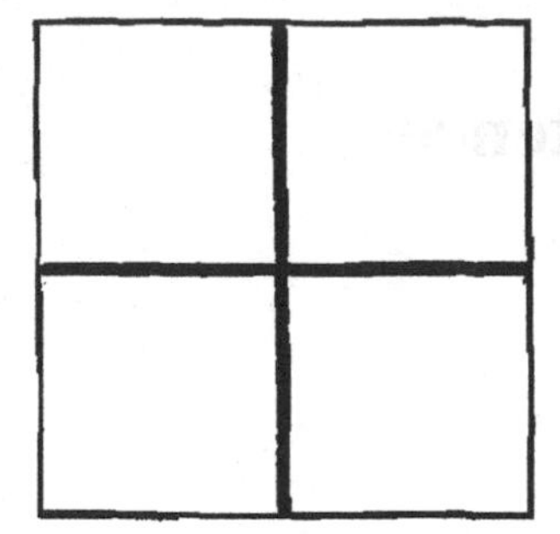

Fig. 7.1

Example 7.2. Prove that among any six points located in a 3×4 rectangle there are at least two points at most $\sqrt{5}$ apart.

Solution. Sometimes we need to allow the pigeonholes to be different in size and shape. In order to solve this problem, let us partition the 3×4 rectangle into five polygons that will be our pigeonholes (Figure 7.2).

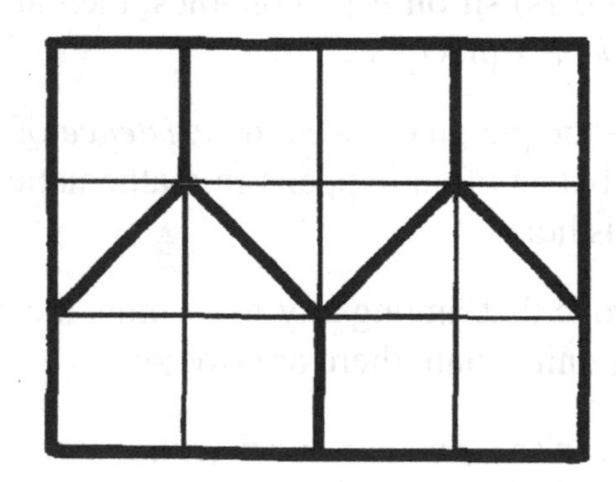

Fig. 7.2 Example of using pigeonholes differing in size and shape

Six pigeons (the given points) sit on five pigeonholes. Therefore at least two pigeons sit on the same pigeonhole. As you can easily compute, the maximal distance between any two pigeons inside the same pigeonhole is $\sqrt{5}$. ■

Exercise 7.2. Prove that no matter how a plane is colored in two colors it must contain two points of the same color exactly one mile apart.

Exercise 7.3. (A. Soifer [S2], [S10]) Prove that among any nine points inside or on the boundary of a triangle of area 1, there are three points that form a triangle of area not exceeding $1/4$.

We can improve the result of the previous exercise:

Exercise 7.4. (A. Soifer [S2], [S10]) Prove that among any seven points inside or on the boundary of a triangle of area 1, there are three points that form a triangle of area not exceeding $1/4$.

The result of Exercise 7.4 can be improved, too; see the book [S2] or its expanded edition [S10].

Exercise 7.5. Suppose all vertices of a convex pentagon lie on the intersections of a grid. Prove that the pentagon (i.e., the interior plus the boundary) contains at least one more intersection of the grid.

Exercise 7.6. (A. Soifer and S. Slobodnik [SS], [S1], [S9]) Forty-one rooks are placed on a 10×10 chessboard. Prove that you can choose five of them that do not attack each other. (We say that one rook *attacks* another if they are in the same row or column of the chessboard.)

Solutions to Exercises

7.1. Assume that there are no pigeonholes that contain $k + 1$ pigeons. Then

$$\begin{array}{c}\text{the 1st hole contains} \leq k \text{ pigeons}\\ \text{the 2nd hole contains} \leq k \text{ pigeons}\\ \\ \underline{\text{the } n\text{th hole contains} \leq k \text{ pigeons}}\\ \text{the total number of pigeons} \leq k \times n\end{array}$$

This contradicts the given fact that there are $kn+1$ pigeons. Therefore, there is a pigeonhole that contains at least $k + 1$ pigeons. ■

7.2. Look at the vertices of an equilateral triangle with side one mile on the colored plane. Since its three vertices (pigeons) are painted in two colors (pigeonholes), we can choose two vertices painted in the same color.

This problem is the starting point of a celebrated open problem. The same statement (i.e., the *existence of two points of the same color distance 1 apart)* can be proven even if the plane is colored in three colors (try to prove it). It is also known that there is a coloring of the plane in seven colors that prevents the existence of two points of the same color 1 unit apart (try to show this too!).

The question is still open for four, five, and six colors after many years and numerous attempts to solve this problem.[1] ■

7.3. Midlines partition the given triangle into four congruent triangles of area $^1/_4$ (Figure 7.3).

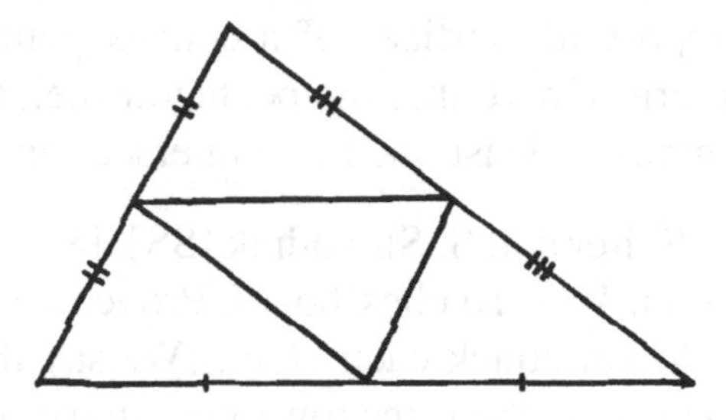

Fig. 7.3

These congruent triangles are our pigeonholes, and the given points are our pigeons. Now nine pigeons are sitting in four pigeonholes. Since $9 = 2 \times 4 + 1$, there is at least one pigeonhole containing at least three pigeons. ■

7.4. Since $7 = 2 \times 3 + 1$, *it would be nice to have three pigeonholes*—then at least one of them would have at least three pigeons! Let us draw only two midlines of the given triangle (Figure 7.4).

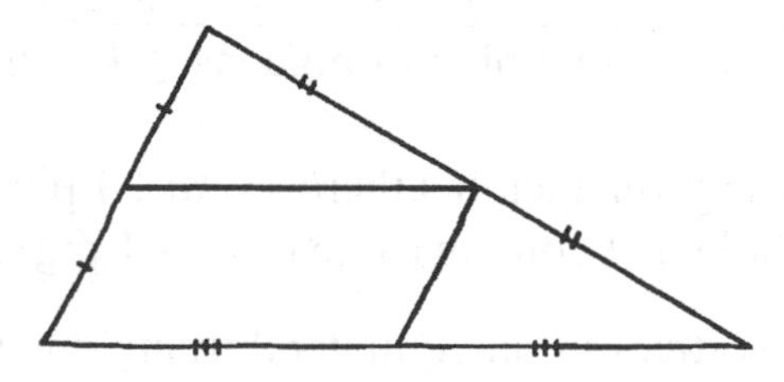

Fig. 7.4

[1] See new to this 2010 Springer edition Chapter 9 dedicated to this problem.

We get three pigeonholes; At least one of them must contain at least three pigeons. If one of the triangles contains three given points, we're done.

If the parallelogram contains three given points, then all we have left to prove is a simple lemma: *The maximum area of a triangle inscribed in a parallelogram of area 1/2 is equal to 1/4.* We leave the proof of this lemma to the reader. ■

7.5. Let us introduce the coordinate system on the grid (Figure 7.5).

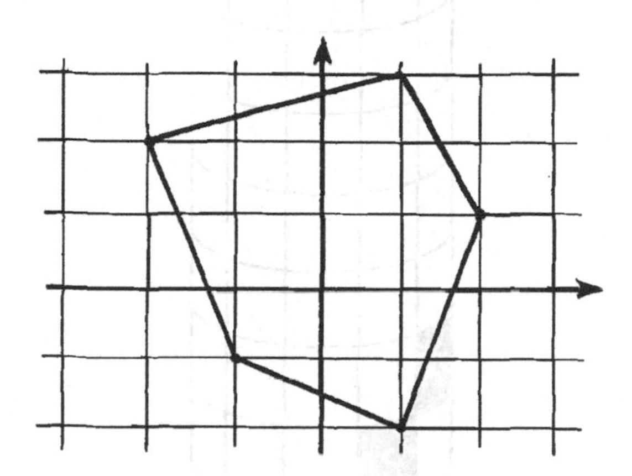

Fig. 7.5 Cartesian coordinate system on a grid

Given a vertex V of a pentagon with the coordinates x, y. We assign to V the ordered pair of the remainders upon division of x, y by 2. There are only four possible outcomes of this operation: (0,0), (0,1), (1,0), and (1,1). These are our pigeonholes. Since we have five pigeons (the vertices of the pentagon), by the Pigeonhole Principle there are two vertices, call them M_1 and M_2, in the same pigeonhole, i.e., their coordinates give the same pair of remainders. In order to complete the proof, all that is left to notice is that the midpoint of the segment M_1M_2 has integral coordinates and lies on the interior or boundary of the pentagon. ■

7.6. This solution was first published in [S1] and also appears in [S9]. Let's make a cylinder out of the chessboard by gluing together two opposite sides of the board. We color the cylinder diagonally in 10 colors (Figure 7.6).

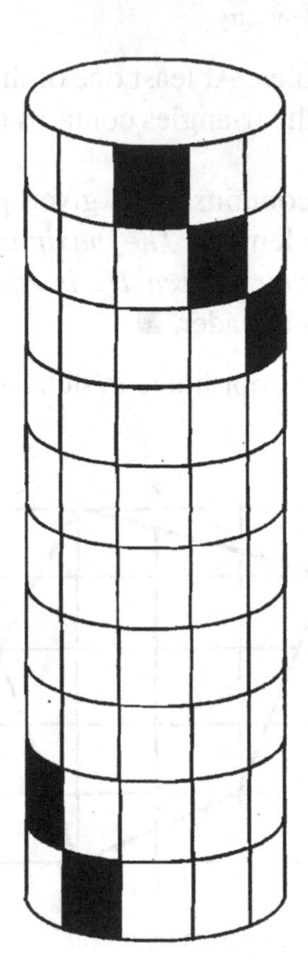

Fig. 7.6 One out of the ten one-color diagonals is shown in black.

Now we have $41 = 4 \times 10 + 1$ pigeons (rooks) in 10 pigeonholes (one-color diagonals). Therefore, there is at least one hole containing at least 5 pigeons. But the 5 rooks located on the same one-color diagonal do not attack each other! ■

8 An Infinite Flock of Pigeons

What would happen if an infinite flock of pigeons were to land on a finite number of pigeonholes? The answer is clear: at least one of the holes would contain an infinite subflock! We will call this simple argument the *Infinite Pigeonhole Principle*. It will allow us to solve a number of important problems.

Let M be an infinite *bounded subset* of the real line R. The word "bounded" means that there exists a positive number m such that $|x| < m$ for every x in M. A point p of R is said to be a *limit point* of the set M if for every $\varepsilon > 0$ the segment $[p - \varepsilon, p + \varepsilon]$ contains infinitely many points of M.

Now we are ready for a classical result of mathematics.

The Bolzano–Weierstrass Theorem 8.1. Every infinite bounded subset M of the real line R has at least one limit point.

Proof. Imagine the points of M as pigeons. So, M is an infinite flock of pigeons. And what do we take as pigeonholes? Since M is bounded, there is a positive integer m such that $|x| < m$ for every x from M. Now we take the segments

$$[-m, -m + 1], [-m + 1, -m + 2], \ldots, [-1, 0], [0, 1], \ldots, [m - 1, m] \qquad (*)$$

as pigeonholes. Then, due to the infinite pigeonhole principle, at least one of them, say, [1, 2], contains infinitely many pigeons. Thus, we have infinitely many pigeons that in decimal form are equal to $1.a_1a_2a_3\ldots$, where $a_1, a_2, a_3, \ldots$ are digits.

The next step is to consider ten segments

$$[1.0, 1.1], [1.1, 1.2], \ldots, [1.8, 1.9], [1.9, 2.0] \qquad (**)$$

as our new pigeonholes. Again we conclude that one of them, say, [1.4, 1.5], contains infinitely many pigeons. Thus, we have infinitely many pigeons of the form $1.4a_2a_3\ldots$

Now from ten segments

$$[1.40, 1.41], [1.41, 1.42], \ldots, [1.49, 1.50] \qquad (***)$$

we choose one, say, [1.43, 1.44], that contains infinitely many pigeons, and so on.

We end up with a real number $x = 1.43\ldots$. We can easily show that x is a limit point of the set M. Indeed, given a positive number ε, we choose an integer m, such that $\frac{1}{10^m} < \varepsilon$. Further, let $x_m = 1.43a_3a_4\ldots a_m$ be a *finite* decimal fraction that we obtain from x by removing all decimal digits of x except the first m digits after the decimal point. Then the segment $[x_m, x_m + \frac{1}{10^m}]$ contains infinitely many pigeons and is contained in the segment $[x - \varepsilon, x + \varepsilon]$. This completes the proof. ■

Let M be a nonempty bounded subset of the real line R. A point α of R is said to be the *exact upper bound* of the set M if it possesses the following properties:

i) there is no point x in M such that $x > a$;
ii) for every positive number ε, the segment $[a - \varepsilon, a]$ contains at least one point of the set M.

The *exact lower bound* of M is defined similarly.

It is clear that if the exact upper bound a of M belongs to M, then a is the *maximal* point of M. If not, then a is a limit point of M. We are ready for another classical theorem of analysis.

Theorem 8.2. *Every nonempty bounded subset of R has the exact upper bound (and the exact lower bound).*

Proof. Let us use again the pigeonholes $(*)$ from the proof of Theorem 8.1. At least one of them contains a point of M. We take the *utmost right* pigeonhole that contains a point of M. Let it be the pigeonhole [1,2]. Then we consider ten pigeonholes $(**)$ and take the utmost right pigeonhole that contains a point of M. Let [1.4,1.5] be this pigeonhole. Then we consider the pigeonholes $(* * *)$ and so on. As a result, we obtain a real number $a = 1.43\ldots$. It can be easily shown (do) that a is the exact upper bound of M. ■

Let now M be an infinite *bounded point set* in the plane R^2. The word "bounded" means that there exists a square (or a disk) that contains M. Here and everywhere in this book by "*disk*" we mean "closed circular disk," i.e., a circle together with all of the points inside it. A point p of R^2 is said to be a *limit point* of M if each disk with center p contains infinitely many points of M.

Plane Sets Theorem 8.3. Any infinite bounded subset M of the plane R^2 has at least one limit point.

Proof. Let m be a positive integer such that for every point $q = (x, y)$ of the set M, the coordinates x, y satisfy the inequalities $|x| < m, |y| < m$. We divide the square with the vertices $(\pm m, \pm m)$ into unit squares (Figure 8.1). These unit squares are the pigeonholes, and

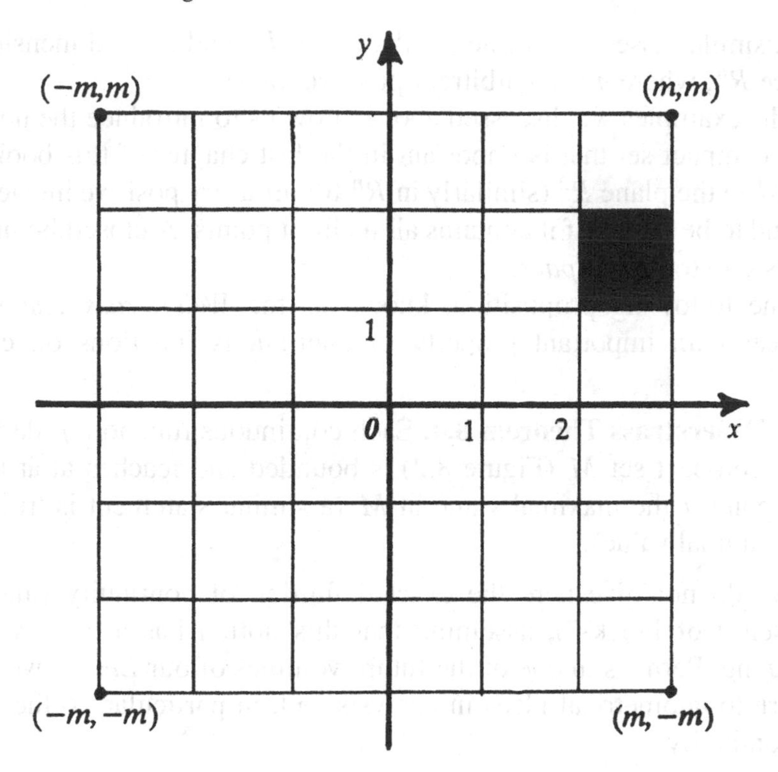

Fig. 8.1

all points of M are the pigeons. By the Infinite Pigeonhole Principle, at least one pigeonhole P contains infinitely many pigeons. Let, for example, the point (2,1) be the bottom left corner of P. Then each pigeon in this pigeonhole is of the form $q = (x, y)$ where $x = 2.a_1a_2a_3\ldots$ and $y = 1.b_1b_2b_3\ldots$.

We now divide the pigeonhole P into 100 new pigeonholes, i.e., 100 small squares with the side length $\frac{1}{10}$. Again, there exists a pigeonhole P_1 that contains infinitely many pigeons. Let, for example, the point (2.7, 1.4) be the bottom left corner of P_1. Then each pigeon in P_1 is of the form $q = (x, y)$ where $x = 2.7a_2a_3\ldots$ and $y = 1.4b_2b_3\ldots$. Now we divide P_1 into 100 pigeonholes that are small squares with the side length $\frac{1}{100}$, and chose a pigeonhole P_2 with the bottom left corner, say (2.76, 1.43), containing infinitely many pigeons, and so on.

We end up with a pair of real numbers $q = (x, y)$ where $x = 2.76\ldots$, $y = 1.43\ldots$. It can be easily shown (do) that q is a limit point of M. We are done. ■

A similar assertion is true for the space R^3 and for n-dimensional space R^n, where n is an arbitrary positive integer.

The examples we discussed above allow us to introduce the notion of a compact set that is important in the last chapter of this book. A set M in the plane R^2 (similarly in R^n for arbitrary positive integer n) is said to be *closed* if it contains all its limit points. A closed bounded set is said to be *compact*.

The following proposition, known as the *Weierstrass Theorem*, indicates an important property of continuous functions on compact sets:

The Weierstrass Theorem 8.4. Each continuous function f defined on a compact set M (Figure 8.2) is bounded and reaches at at least one point a the maximal value in M (a similar statement is true for the minimal value).

We do not give here the exact definition of continuity (that is, "absence of breaks"), assuming that this notion has a clear visual meaning. Perhaps in one of the future volumes of our *Etudes* we will return to geometrical ideas in analysis and, in particular, to the idea of continuity.

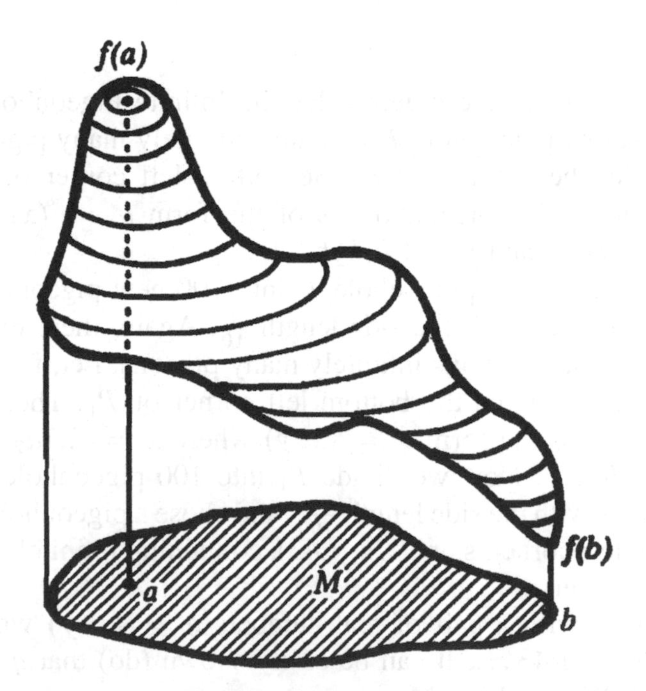

Fig. 8.2 Queen Dido's Puzzle

The following proposition, known as the *Intermediate Value Theorem,* expresses another important property of continuous functions.

Intermediate Value Theorem 8.5. Let M be a connected set (that is, M consists of "one piece") and f be a continuous function on M. If a and b are two points of M and y is a number such that $f(a) < y < f(b)$, then there exists a point c of M for which $f(c) = y$ (Figure 8.3).

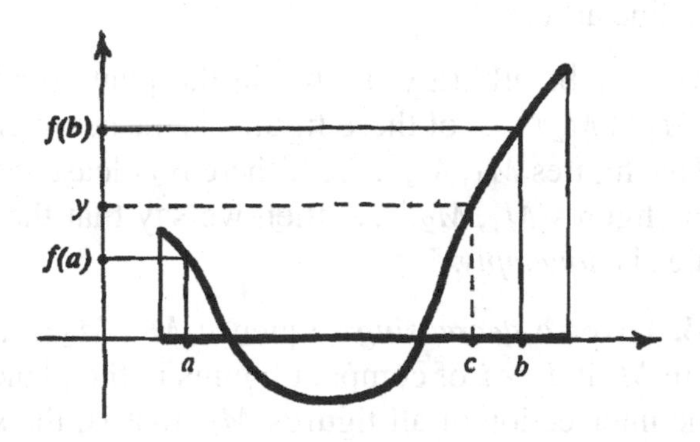

Fig. 8.3

Exercise 8.1. A compact figure F of area S and a line L are given in the plane. Prove that there exists a line parallel to L that divides F into two parts, each of area $\frac{1}{2}S$ (Figure 8.4).

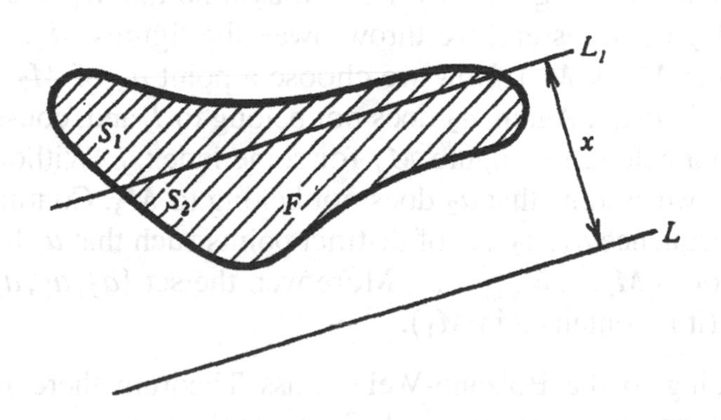

Fig. 8.4

Exercise 8.2. A compact figure F of area S and a point p are given in the plane. Prove that there exists a line through p that divides F into two parts, each of area $\frac{1}{2}S$.

Exercise 8.3. The Russian mathematician Pavel Uryson noticed that for every compact figure F in the plane there exists a square circumscribed about F. Can you prove it?

The above exercises and similar problems can be found in [S2], [S10] and the fine article [T].

Let $M_1, M_2, \ldots$ be arbitrary figures in the plane (or in R^n). The *intersection* $M_1 \cap M_2 \cap \ldots$ of these figures is the set of all points that belong to *all* the figures $M_1, M_2, \ldots$. If there is at least one point common to all the figures $M_1, M_2, \ldots$, then we say that the intersection of these figures is *nonempty*.

Example 8.4. Given a *decreasing* sequence $M_1, M_2, \ldots$ that is, M_j is contained in M_i if $j > i$ of compact figures in the plane (or in R^n). Prove that the intersection of all figures M_i (that is, the set of points that belong to all the figures) is a nonempty compact figure.

Proof. The intersection of closed sets is always closed. The intersection of bounded figures is bounded. So, it remains to prove that the intersection I of a decreasing sequence of nonempty compact figures M_i is nonempty. Assume the opposite, i.e., that is empty. We take a point a_1 of the figure M_1. Then a_1 does not belong to I (since I is empty). So there exists an index i such that a_1 does not belong to M_i. Without loss of generality we can assume that a_1 does not belong to M_2 (if necessary, we throw away the figures $M_2, \ldots, M_{i-1}$ and rename M_i as M_2). Now we choose a point a_2 of M_2. Then a_1 is distinct from a_2. Again, a_2 does not belong to I and, consequently, a_2 does not belong to a figure M_j for some index j. Without loss of generality we assume that a_2 does not belong to M_3. Continuing, we obtain a sequence $a_1, a_2, \ldots$ of distinct points such that a_i belongs to M_i but not to $M_{i+1}, M_{i+2}, \ldots$. Moreover, the set $\{a_1, a_2, a_3, \ldots\}$ is bounded (it is contained in M_1).

According to the Bolzano-Weierstrass Theorem there is a limit point b of the set $\{a_1, a_2, a_3, \ldots\}$. Since all the points $a_1, a_2, a_3, \ldots$, except $a_1, a_2, a_3, \ldots, a_{i-1}$, belong to M_i, b also belongs to M_i (recall that M_i is closed, that is, it contains all its limit points). Thus, each figure M_i $(i = 1, 2, \ldots)$ contains b, contradicting the assumption that the intersection of all these figures is empty. ■

Let us note that the theorem we just proved is not true for noncompact figures (even closed). For example, let us denote the half-plane determined in a coordinate system (x, y) by the inequality $x \geq i$ (where $i = 1, 2, \ldots$) by M_i. We obtain a decreasing sequence of closed noncompact figures $M_1, M_2, \ldots$, such that their intersection is empty (see Figure 18.14). We are done.

Now let us introduce a new notion. Let M be a figure in the plane, and ε a positive number. A point x is said to be *ε-close* to the figure M, if there exists a point y of M such that the distance $|xy|$ between these points does not exceed ε. The set $E_\varepsilon(M)$ of all points x that are ε-close to M is called the *ε-extension* of the figure M (Figure 8.5). In other words, the ε-extension of M is the union of all disks of radius ε with centers at the points of the figure M. It can be easily shown that if M is compact, then its ε-extension is also compact.

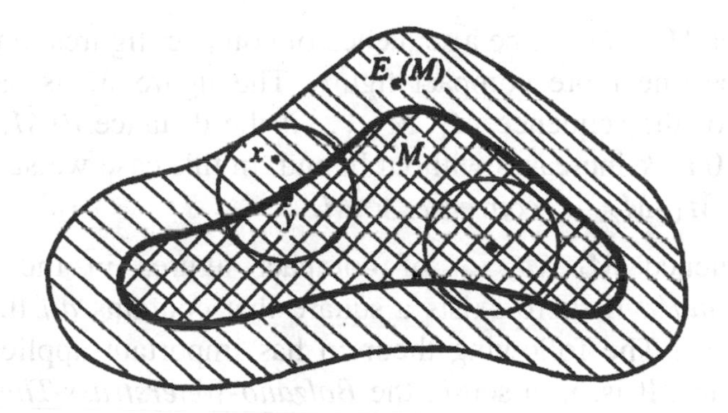

Fig. 8.5

The German mathematician Felix Hausdorff introduced a notion of *distance between two compact figures*. Here is his definition: let M_1 and M_2 be two distinct compact figures in the plane. The *distance d* (M_1, M_2) between M_1 and M_2 is the least positive number ε such that M_1 is contained in the ε-extension of M_2 and M_2 is contained in the ε-extension of M_1.

For example, if M_1 and M_2 are two equilateral triangles with parallel sides and common centroid (Figure 8.6), then the distance $d(M_1, M_2)$ is equal to $h\sqrt{3}$, where h is the distance between the cor-

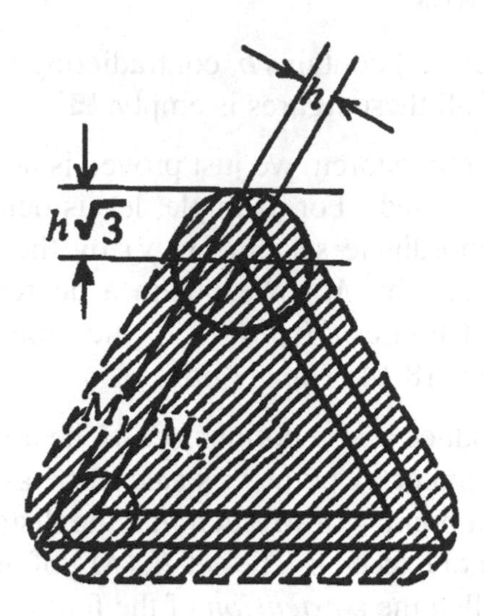

Fig. 8.6

responding parallel sides of the triangles. Indeed, the h-extension of the smaller triangle does not contain the vertices of the larger one.

Now let $M_1, M_2, \ldots$ be a sequence of compact figures in the plane and M be one more compact figure. The figure M is said to be the *limit* of the sequence $M_1, M_2, \ldots$ if the distance $d(M, M_k)$ approaches 0 as k increases without bound. In this case we say that the sequence $M_1, M_2, \ldots$ *converges* to M.

A sequence $M_1, M_2, \ldots$ of compact figures in the plane is called *bounded* if there exists a square that contains *all* the figures $M_1, M_2, \ldots$. The following theorem has important applications in mathematics. It is, in a sense, the *Bolzano-Weierstrass Theorem for Compact Figures*.

Bolzano-Weierstrass Theorem for Compact Figures 8.6. Any bounded sequence of compact figures in the plane (or more generally, in the space R^n) has a convergent subsequence.

Proof. Have you forgotten the pigeonholes? Let $M_1, M_2, \ldots$ be a bounded sequence of compact figures, and m be a positive integer such that all the figures $M_1, M_2, \ldots$ are contained in the square with vertices $(\pm m, \pm m)$. We divide this square into unit squares (Figure 8.1) and presume that each unit square contains its boundary. Now each union of any number of unit squares we call a pigeonhole.

So, there is a finite number of pigeonholes (more precisely, there are $4m^2$ unit squares and, consequently, 2^{4m^2} pigeonholes, including the empty pigeonhole, that is, the "union" of the empty set of squares). Certainly, our compact sets $M_1, M_2, \ldots$ are the pigeons! But in what sense is a pigeon sitting on a pigeonhole?

Let F be a compact figure contained in the square with the vertices $(\pm m, \pm m)$. We denote the union of the unit squares that have at least one common point with F by $C_1(F)$. The figure $C_1(F)$ is said to be the *container* of F.

Now we say that the pigeon M_k is sitting on a pigeonhole if this pigeonhole is the container of M_k. So, each pigeon is sitting on *only one* pigeonhole. The pigeon M_k and the pigeonhole on which it is sitting are drawn in Figure 8.7.

Thus, we have a finite number of pigeonholes and infinitely many pigeons. Therefore, by the Infinite Pigeonhole Principle, there exists a pigeonhole P_1 that contains infinitely many pigeons. We pick one of

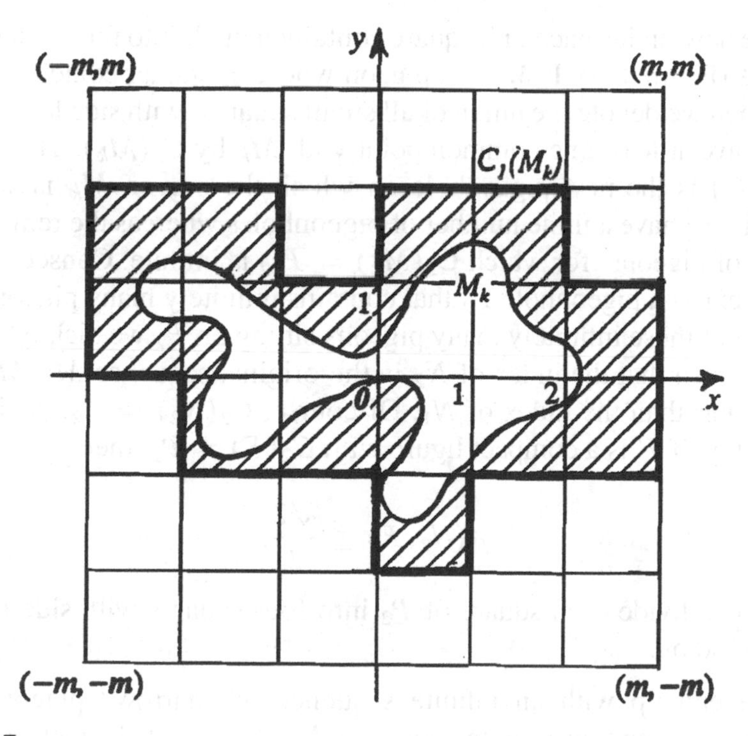

Fig. 8.7

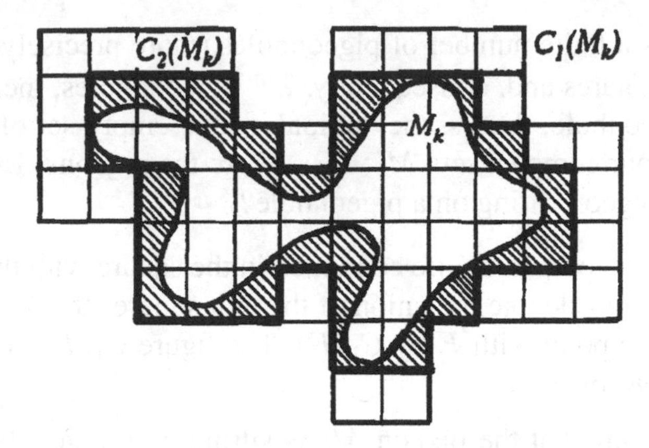

Fig. 8.8

these pigeons and denote it by N_1. Obviously, $C_1(N_1) = P_1$. It is not difficult to show that if F is a compact figure with $C_1(F) = P_1$, then

$$d(N_1, F) \leq \sqrt{2},$$

since the length of the diagonal of the unit square is equal to $\sqrt{2}$.

We now divide each unit square contained in P_1 into four squares of side $\frac{1}{2}$ (Figure 8.8). If M_k is a pigeon whose container coincides with P_1, then we denote the union of all small squares (with side length $1/2$) that have at least one common point with M_k by $C_2(M_k)$. The figure $C_2(M_k)$ is the new pigeonhole on which the pigeon M_k is sitting. Again, we have a finite number of pigeonholes whereas the remaining flock of pigeons (for which $C_2(M_k) = P_1$) is infinite. Consequently, there exists a pigeonhole P_2 that contains infinitely many pigeons.

Out of these infinitely many pigeons sitting on P_2 we pick one, call it N_2, such that the index of N_2 in the original sequence $M_1, M_2, \ldots$ is greater than the index of N_1. Of course, $C_2(N_2) = P_2$. As in the first step, if F is a compact figure with $C_2(F) = P_2$, then

$$d(N_2, F) \leq \frac{\sqrt{2}}{2}.$$

We then divide each square of P_2 into four squares with side length $\frac{1}{4}$, and so on.

We end up with an infinite sequence of "narrow" pigeonholes $P_1, P_2, \ldots,$ and an infinite sequence $N_1, N_2, \ldots$ of distinct pigeons that is a subsequence of the given sequence $M_1, M_2, \ldots$ of compact figures. According to the construction, for each pigeon N_k of the created subflock, we have $C_1(N_k) = P_1$; for each N_k with $k \geq 2$, we have $C_2(N_k) = P_2$, and so on. This means that

$$d(N_1, N_k) \leq \sqrt{2} \text{ for all } k > 1,$$

$$d(N_2, N_k) \leq \frac{\sqrt{2}}{2} \text{ for all } k > 2,$$

and generally,

$$d(N_j, N_k) \leq \frac{\sqrt{2}}{2^{j-1}} \text{ for all } k > j.$$

Finally, we denote the intersection of all pigeonholes $P_1, P_2, \ldots$ by N. Then N is a nonempty compact figure (see Example 8.4). It can be easily shown that $C_1(N) = P_1, C_2(N) = P_2, \ldots$. Consequently, $d(N_j, N) \leq \frac{\sqrt{2}}{2^{j-1}}$. This means that the sequence $N_1, N_2, \ldots$ converges to N. This completes the proof. ■

Example 8.5. Mathematicians of ancient Greece knew that *among all figures of given perimeter P the circle has maximal area*. They knew it but could not prove it! A nice proof was suggested in the nineteenth century by the Swiss mathematician Jacob Steiner. But his proof had a hole. We discuss here the wonderful idea of Steiner and fill in the hole in his proof. This will serve as a good example of an application of the above Theorem 8.4. The problem of determining the figure F of maximal area with the given perimeter P is called the *isoperimetric problem*. In this case, the figure F is called *extremal*.

Let F be a figure with the given perimeter P. If it is not convex, then we slide a rubber loop over it (Figure 8.9). We obtain a figure

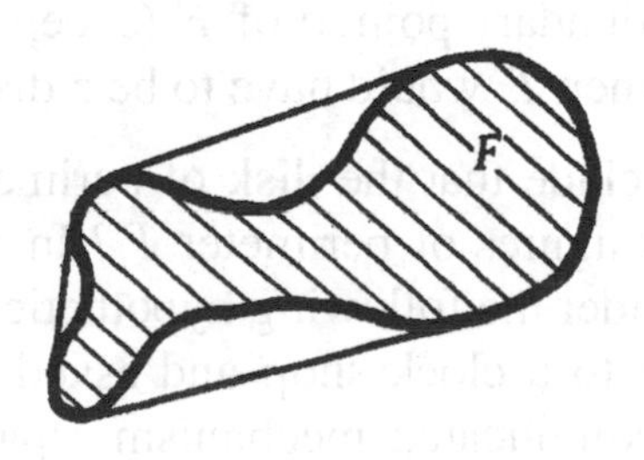

Fig. 8.9

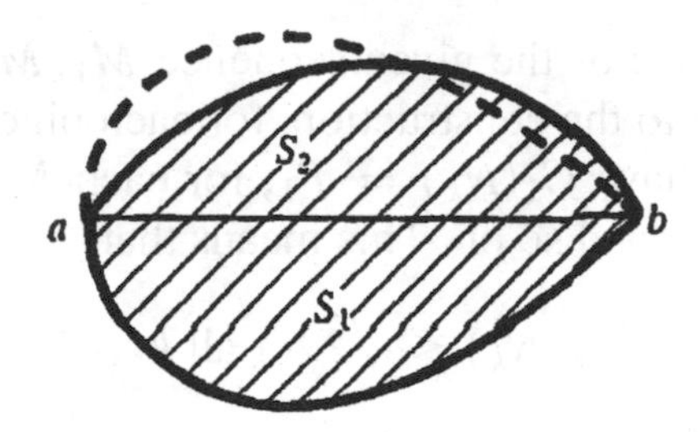

Fig. 8.10

of a larger area and smaller perimeter (we'll discuss convex figures in greater detail in Chapter 4). So, if a figure is not convex, then it is not extremal, that is, it cannot be a solution of the isoperimetric problem.

Let F be a convex figure. We call a chord $[a, b]$ of F a *cross-cut* if it divides the boundary of F into two arcs of equal length $\frac{1}{2}P$. Steiner noticed that if F is extremal, then every cross-cut divides its area into two *equal* parts. Indeed, if *ab* is a cross-cut and the areas of parts S_1 above and S_2 below it are unequal and $|S_1| > |S_2|$ (Figure 8.10), then by replacing S_2 with the symmetric image of S_1, we obtain a figure with the same perimeter P and a *greater* area, contradicting the extremality of F.

Steiner fixed a cross-cut $[a, b]$ of the extremal figure F and showed that for any boundary point c of F distinct from a and b, the angle *acb* must be equal to 90°. Indeed, if the angle *abc* were not equal to 90° (Figure 8.11), then we could install a hinge in the point c and rotate one of the shaded areas about c until the angle *acb* is equal to 90° (Figure 8.11). While doing so, we certainly would not change the shaded areas, but would increase the area of the triangle *acb* (prove it). Now we can replace the lower half of F with a symmetric image of the upper half. Thus, we get a figure of the same perimeter as F but of a greater area.

But if for every boundary point c of F (except a and b), the angle *acb* is equal to 90°, then F would have to be a disk.

May we now conclude that the disk of perimeter P has the maximal area among all figures of perimeter P? In order to answer this question, let us consider the following hypothetical situation.

A customer came to a clock shop and asked to have his antique clock with a very complicated mechanism repaired. The owner of the shop asked his experts for help, and they replied that nobody

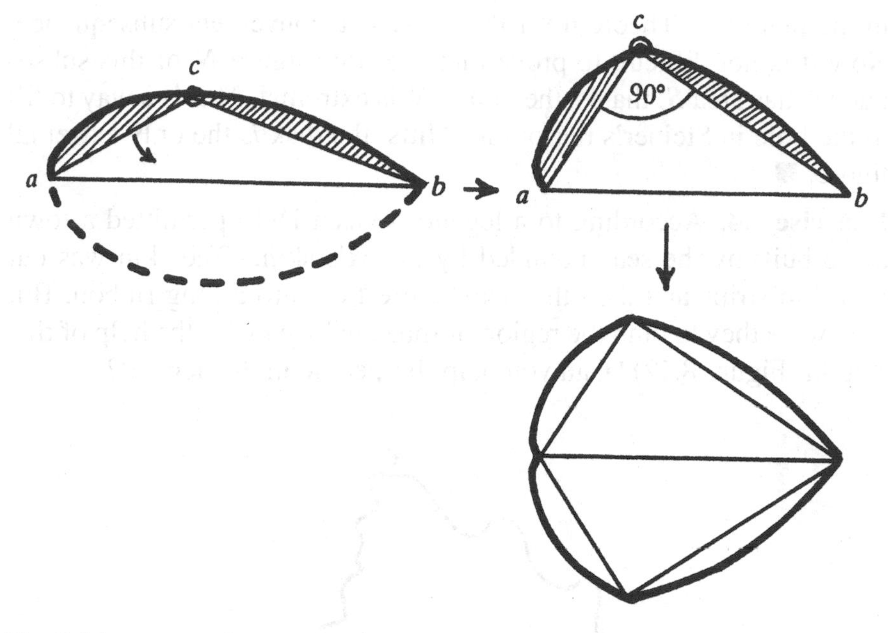

Fig. 8.11

except the foreman Smith could repair the clock. Smith was ill so the customer was asked to come next week. However, the next week it became clear that the foreman Smith also was unable to repair that clock. So "nobody except Smith can" does not mean yet that "Smith can."

Similarly, Steiner's reasoning above shows that no figure except the disk can be extremal. Does it mean that the disk is surely the extremal figure? Not at all! Perhaps no figure except the disk is extremal, and the disk is not extremal either. If we had a guarantee of the *existence* of an extremal figure, then (knowing that no figure except the disk is extremal) we could conclude that the disk *really is* the extremal figure. The lack of such a guarantee of the existence was the hole in Steiner's reasoning.

And how can we get a guarantee for the existence of an extremal figure? Theorem 8.4 gives such a guarantee! Indeed, let us denote the exact upper bound of areas of all figures with perimeter P by S. Then for every positive integer k there exists a figure M_k of perimeter P whose area is greater than $S - \frac{1}{k}$. We can place all the figures $M_1, M_2, \ldots$ in a bounded piece of the plane (say, in a circle of radius P). We obtain the bounded sequence $M_1, M_2, \ldots$ of compact figures in the plane. By Theorem 8.1 there exists a convergent subsequence. Now it is not difficult to prove that the limit figure N of this subsequence has area S, that is, the figure N is extremal. This is a way to fill in the hole in Steiner's reasoning. Thus, the disk *is* the only extremal figure. ■

Exercise 8.4. According to a legend, Queen Dido permitted a town to be built by the sea "bounded by an ox's skin." The skin was cut into thin strips and then the strips were tied into a long ribbon. But how were they to bind the region of maximal area with the help of this ribbon (Figure 8.12)? Can you help the people in this legend?

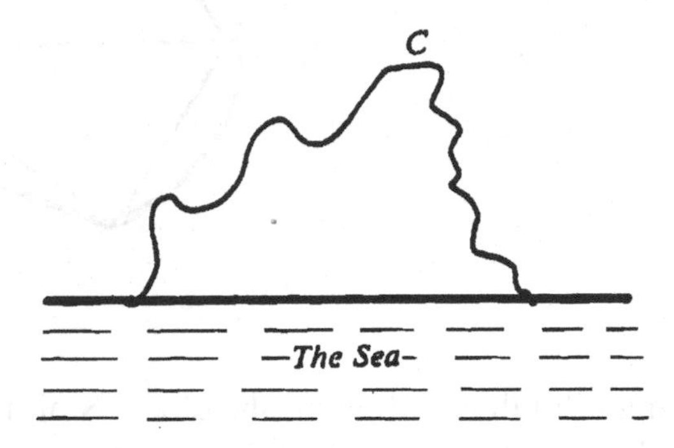

Fig. 8.12 Queen Dido's Puzzle

Exercise 8.5. (B. Grünbaum; same as Exercise 6.9) Prove that every tiling of an infinite strip by copies of a polyominal tile T contains a bounded part that can be used to tile a cylinder.

Solutions to Exercises

8.1. Let us denote by x the distance between the given line L and the constructed line L_1 that cuts F into two parts (Figure 8.4). We denote by S_1 and S_2 the areas of these parts. Then $S_1 - S_2$ is a continuous function of x (we don't give a proof of this visually clear assertion because we did not introduce the exact definition of continuity). But when x is small, $S_1 - S_2$ is equal to S (Figure 8.4), and when x is large $S_1 - S_2 = -S$. So, according to the Intermediate Value Theorem, there is a value $x = c$ for which $S_1 - S_2 = 0$, that is, $S_1 = S_2 = \frac{1}{2}S$. ■

8.2. Let L be a directed line through p that forms angle ϕ with a fixed initial ray L_0 (Figure 8.13). We denote the areas of the parts into which L divides the figure F by S_1 and S_2 (S_1 is to the left of L). Then $S_1 - S_2$ is a continuous function of ϕ. But as ϕ runs through all the values from 0 to π, the areas S_1 and S_2 interchange their roles. So, if $S_1 - S_2$ is negative for $\phi = 0$, then it will be positive for $\phi = \pi$. Consequently, by the Intermediate Value Theorem, there exists a value $\phi = c$ for which $S_1 - S_2 = 0$, that is $S_1 = S_2 = \frac{1}{2}S$. ■

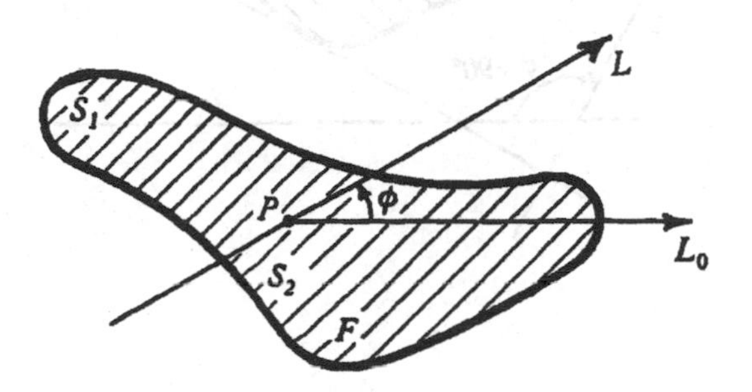

Fig. 8.13

8.3. We squeeze the figure F between two parallel lines L_1 and L_2 that form angle ϕ with a fixed initial ray (Figure 8.14). Then we squeeze F between two parallel lines M_1 and M_2 that form the angle $\phi + 90°$ with the initial ray (Figure 8.15). The four lines define a rectangle circumscribed about F. Let a be the length of the side parallel

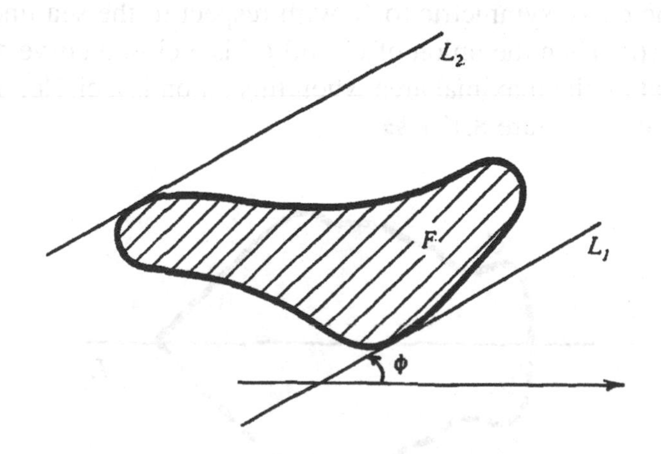

Fig. 8.14

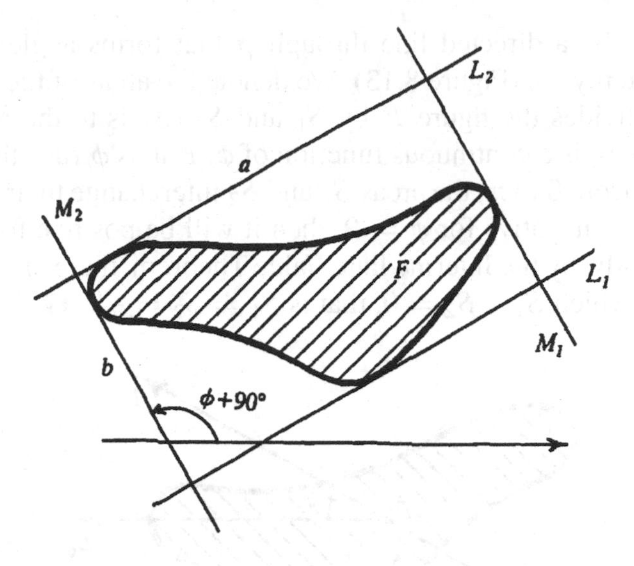

Fig. 8.15

to L_1 and b be the length of the side parallel to M_1. Then $a - b$ is a continuous function of ϕ. But when ϕ runs through the values from 0 to π, the lengths a and b interchange their roles. This means that if $a - b > 0$ for $\phi = 0$, then $a - b < 0$ for $\phi = \pi$. Consequently, there exists an angle $\phi = \mathrm{c}$ for which $a - b = 0$, that is, the circumscribed rectangle turns into a square. ■

8.4. Let C be a curve bounding the town (Figure 8.12), q its length, and C' the curve symmetric to C with respect to the sea line L (see Figure 8.16). Then the union of C and C' is a closed curve of length $2q$. We obtain the maximal area when this union is a circle; hence, C is a semicircle (Figure 8.17). ■

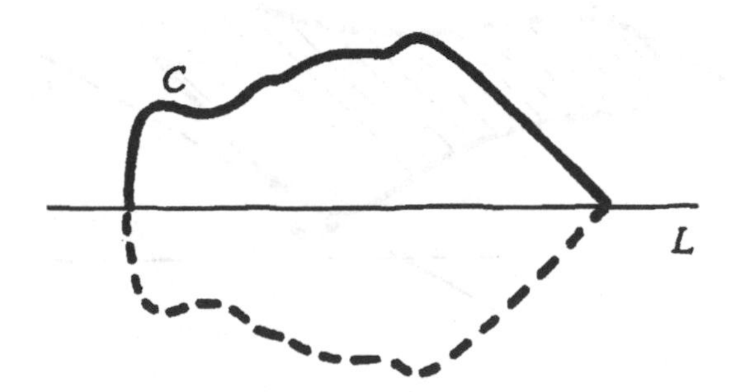

Fig. 8.16

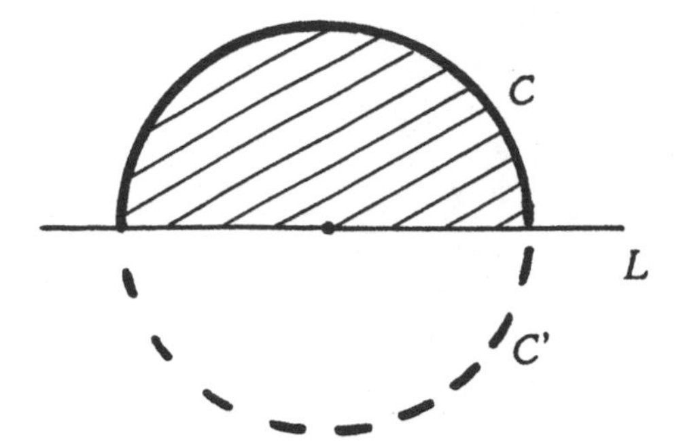

Fig. 8.17

8.5. Let S be an infinite strip tiled by copies of tile T. There are finitely many shapes of step-lines (can you prove it?) cutting across S along the boundaries of tiles (Figure 8.18 shows one such step-line cut).

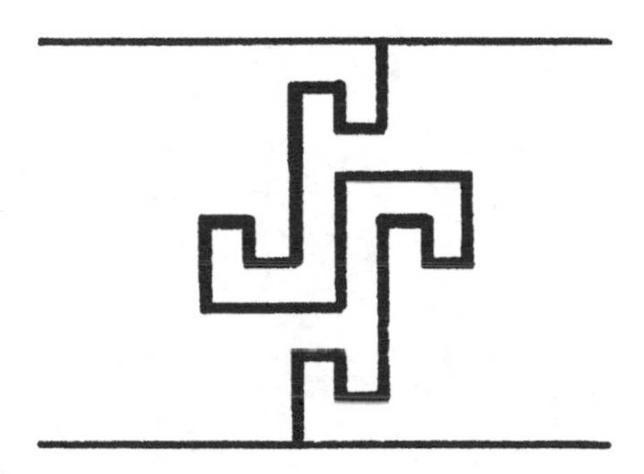

Fig. 8.18

On the other hand, there are infinitely many step-line cuts because S is an infinite strip. By the Infinite Pigeonhole Principle, there is a cut that repeats at least twice. We get a region F just like the one in Figure 6.5; you can glue a cylinder out of F. ■

Chapter 3
A Word About Graphs

9 Combinatorics of Acquaintance, or an Introduction to Graph Theory

In mathematics, it is sometimes possible to derive something out of seemingly nothing, as the following problem illustrates.

Example 9.1. A number of people (more than one) come to a party. Prove that at least two of them have an equal number of acquaintances at the party. (The notion of acquaintance is reflexive: if A is acquainted with B, then B is acquainted with A.)

Solution. If n people come to the party, then for each person the number of acquaintances is an integer ranging from 0 to $n - 1$. In fact, 0 and $n - 1$ may not both serve as numbers of acquaintances for people in our party because 0 implies the existence of a person not acquainted with anybody and $n - 1$ implies the existence of a person acquainted with everyone.

Thus, we have $n-1$ possibilities for number of acquaintances and n people in the party. The Pigeonhole Principle now proves the required result. ■

In the solution to Example 9.1 we discovered that at most $n - 1$ integers can appear as numbers of acquaintances: $0, 1, \ldots, n - 2$ or $1, 2, \ldots, n - 1$. Curiosity prompts the following question.

Exercise 9.1. Is there a party of n such that

(a) every integer $0, 1, \ldots, n - 2$ appears as the number of acquaintances of a person at the party?
(b) every integer $1, 2, \ldots, n - 1$ appears as the number of acquaintances of a person at the party?

A. Soifer, *Geometric Etudes in Combinatorial Mathematics*,
DOI 10.1007/978-0-387-75470-3_3,

Exercise 9.2. Can it so happen that every person has exactly 7 acquaintances in a party of 21?

If we know that at least six people are at a party, we can prove another exciting result, which is a simple particular case of the powerful theorem proven by the English mathematician Frank Plumpton Ramsey in 1928 and published posthumously in 1930.

Example 9.2. Prove that in any party of six people there are three mutual acquaintances or three mutual non-acquaintances.

Proof.

(a) It is convenient to record the information about acquaintances of a group of six people with a combination of two diagrams, G and $\overline{G}$. In both diagrams we represent ach person by a vertex for a total of six vertices, and two vertices of G are connected by a curve (the curve's shape is irrelevant) if and only if they correspond to two people who are acquainted. Two vertices of $\overline{G}$ are connected by a curve if and only if they correspond to two people who are not acquainted.

Note that any two vertices are connected in exactly one of the diagrams G, $\overline{G}$.

The given problem is now equivalent to proving that at least one of the two diagrams G, $\overline{G}$ contains a triangle (i.e., three pairwise connected vertices)!

(b) Let us fix the vertex A in both G and $\overline{G}$. A is connected to each of the five other vertices either in G or in $\overline{G}$; therefore, it must be connected to at least three of the five vertices in G or in $\overline{G}$ (Pigeonhole Principle with five pigeons and two holes).

Due to the symmetry of the problem, we can assume without loss of generality that A is connected to the vertices B_1, B_2, and B_3 in G. If any two of the vertices B_1, B_2, B_3 are connected in G, then these two vertices and A are the vertices of a triangle contained in G (Figure 9.1).

If no two of the vertices B_1, B_2, B_3 are connected in G, then the triangle B_1, B_2, B_3 is contained in $\overline{G}$. ■

There is another way to look at the party-of-six problem. Given a regular hexagon with every two of its vertices connected by an edge the so-called complete graph on 6 vertices (Figure 9.2).

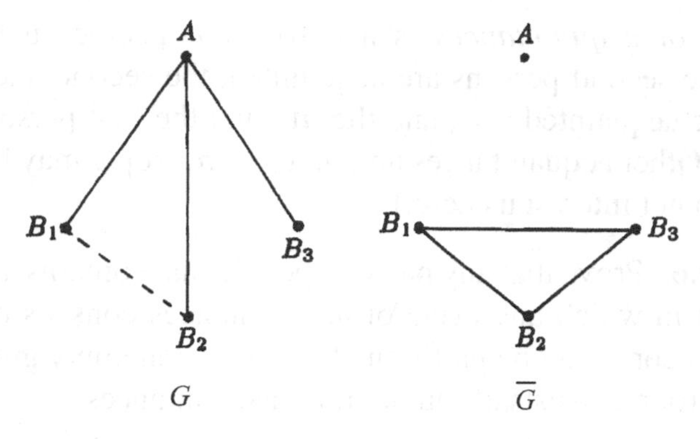

Fig. 9.1 The Party of Six Puzzle

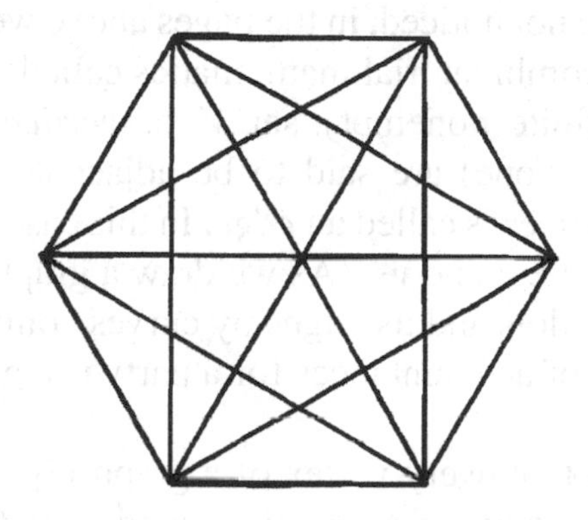

Fig. 9.2 Complete graph on 6 vertices

The party-of-six statement is equivalent to the following: *No matter how the edges of the complete graph with 6 vertices are colored in two colors (one color per edge), there is always a one-color triangle (with vertices in the given points).*

Exercise 9.3. Prove that six is the smallest number of people at the party to guarantee the result of Example 9.2, i.e., a party of five people might have neither three mutual acquaintances nor three mutual non-acquaintances.

Exercise 9.4. Prove that in any party of nine there are three mutual acquaintances or four mutual non-acquaintances.

Exercise 9.5. Prove that nine is the smallest number of people in the party to guarantee the result of Exercise 9.4.

A *cycle of acquaintances* is a party of n people such that the first and the second persons are acquainted, the second and the third persons are acquainted, ..., and the nth and the first persons are acquainted. (Other acquaintances among these n people may be present, but they do not interest us here.)

Exercise 9.6. Prove that any party of people that contains at least two people and in which any cycle of acquaintances consists of an even number of people can be partitioned into two nonempty groups, such that each group consists of mutual non-acquaintances.

Exercise 9.7. Prove the converse of the statement of Exercise 9.6.

In case you have not noticed, in the pages above we have introduced you to an area of combinatorial mathematics called *Graph Theory*.

A *graph* is a finite nonempty set V of *vertices*, some pairs of which (or perhaps none) are said to be adjacent. An adjacent pair $e = \{v_1, v_2\}$ of vertices is called an *edge*. In this case we say that e and v_1 are *incident*, as are e and v_2. (As we draw a graph, we often represent its vertices by dots and its edges by curves connecting the dots.)

The diagram G of acquaintances for a party is a good example of a graph.

If it so happens that every vertex of a graph G_1 is also a vertex of graph G, *and* every edge of G_1 is also an edge of G, then the graph G_1 is said to be a *subgraph* of the graph G.

The diagram $\overline{G}$ of non-acquaintances is uniquely determined by the graph of acquaintances G, and is called the complementary graph of G or simply the *complement* of G.

The number of edges incident to a vertex v in a graph is called the *degree* of v and is denoted deg v. The following theorem can be proved using identical reasoning to Exercise 9.1. The symbol $\sum_{v \in V} \deg v$ in Theorem 9.1 denotes the sum of all numbers deg v where $v \in V$ (i.e., v runs through all the elements of the set V).

Theorem 9.1. *Prove that for any graph G*

$$\sum_{v \in V} \deg v = 2q,$$

where V is the set of vertices of G, and q is the number of edges.

If the vertex set V of G can be partitioned into two non-empty subsets V_1 and V_2 such that no edge connects vertices in either of the subsets (but may connect vertices from V_1 to vertices from V_2), then F is called a *bipartite graph*.

Exercises 9.6 and 9.7 provide a nice description of bipartite graphs. In the language of graph theory it is as follows:

Theorem 9.2. *A graph G is a bipartite if and only if it contains no odd cycles.*

Let G be a bipartite graph g with the vertex set V partitioned into subsets V_1 and V_2 with n_1 and n_2 elements respectively. If *every* vertex of V_1 is adjacent to *every* vertex of V_2, then G is called a *complete bipartite graph* and denoted K_{n_1,n_2}. Obviously, K_{n_1,n_2} contains $n_1 n_2$ edges. Figure 9.3 gives examples of all complete bipartite graphs with six vertices.

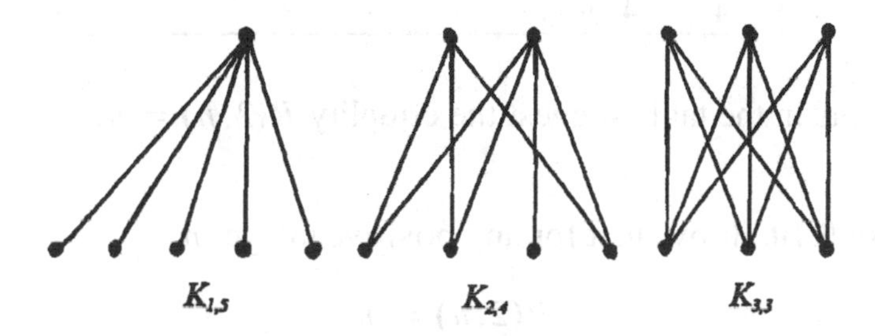

Fig. 9.3 Complete bipartite graphs on 6 vertices

Exercise 9.8. Let $n = n_1 + n_2$ and $n = m_1 + m_2$ be two distinct decompositions of n into the sum of two positive integers (i.e., $n_1 \neq m_1$ and $n_1 \neq m_2$). Is there a graph G such that $G = K_{n_1,n_2}$ and at the same time $G = K_{m_1,m_2}$?

A graph corresponding to a party of n mutual acquaintances is called a *complete graph* and is denoted K_n.

For positive integers m and n, the *Ramsey number* $R(m,n)$ is the smallest positive integer R such that for any graph G with R vertices, G contains K_m as a subgraph or $\overline{G}$ contains K_n as a subgraph.

Now we are ready to translate Example 9.2 and Exercises 9.2, 9.3, and 9.4 into the language of graph theory or, more precisely, *Ramsey Theory* — you will find much about this relatively new field in my *Mathematical Coloring Book* [S7].

Theorem 9.3. $R(3,3) = 6$; $R(3,4) = 9$.

Of course, the function $R(m,n)$ is symmetric:

Exercise 9.9. For any positive integers m and n

$$R(m,n) = R(n,m).$$

Now, nearly eighty years after the pioneering paper by Frank P. Ramsey, we still know only a few small Ramsey numbers. They all are presented in the following table (Table 9.1):

Table 9.1 Ramsey Numbers

$n \backslash m$	2	3	4	5	6	7	8	9	...
2	2	3	4	5	6	7	8	9	...
3	3	6	9	14	18	23		36	
4	4	9	18						

The dots in the table suggest the equality $R(2,n) = n$.

Exercise 9.10. Prove that for any positive integer n,

$$R(2,n) = n.$$

But even $R(4,5)$ or $R(5,5)$ are not known. So if you find new partial solutions to the following problem be sure to write to us![1]

Open Problem 9.1. For positive integers m and n find

$$R(m,n).$$

This is an extremely difficult problem because of its generality. Any partial results would be most welcome.[2]

[1] See Chapter 7 of this book for great developments that have taken place since the first edition was published in 1991.

[2] See the new in this edition Chapter 7 for more small Ramsey numbers.

Solutions to Exercises

9.1. (a) We can prove by induction on $n$that there is a party of n people such that every integer $0, 1, \ldots, n-2$ appears as the number of acquaintances of a party member.

Two non-acquainted people deliver an example of such a party for $n = 2$.

Assume that there is a party P of n people such that every integer $0, 1, \ldots, n-2$ appears as the number of acquaintances. In fact, by the Pigeonhole Principle, one of these numbers, say $k(0 \leq k \leq n-2)$, must appear twice.

Now we have to construct a party P' of $n+1$ people such that every integer $0, 1, \ldots, n-1$ appears as the number of acquaintances. We start with the party P of n people and add one more person, who is acquainted with exactly one of two people with k acquaintances, and with everyone having more than k acquaintances. You can easily verify that P' satisfies the required condition.

(b) There is such a party. The construction is similar to the one in Exercise 9.1(a). ■

9.2. Due to Theorem 9.1, the total number T of acquaintances is even, on the other hand, if we assume that there exists a party of 21 such that everyone has exactly seven acquaintances, we have $T = 21 \times 7$, an odd number. This contradiction proves that such a party does not exist.

We would like to mention here that similar reasoning proves the following result: *there is no polyhedron with an odd number of odd-sided faces.* ■

9.3. Behold (Figure 9.4):

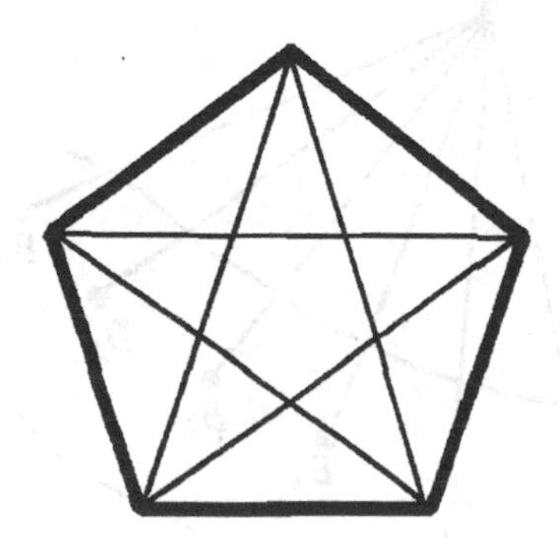

Fig. 9.4

9.4. Assume there is a person, A acquainted with at least four party members, say, B_1, B_2, B_3, and B_4 (Figure 9.5).

Then if at least two of the B_i, say, B_1 and B_2, are acquainted, we get a triangle AB_1B_2 of mutual acquaintances. Otherwise B_1, B_2, B_3, and B_4 are mutually non-acquainted.

Now consider the opposite case, i.e., every person is acquainted with at most three party members. This means that everyone has at least five non-acquaintances. Now, there can not be a party of nine such that every party member has exactly five non-acquaintances (because the total number of non-acquaintances must be even, whereas 9×5 is odd — reasoning similar to the solution of Exercise 9.2). Therefore, at least one party member, say, A, has at least six non-acquaintances $B_1, B_2, \ldots, B_6$ (Figure 9.6).

All that is left to do is to apply the party-of-six result (Example 9.2) to the party of $B_1, B_2, \ldots, B_6$. If three of them are

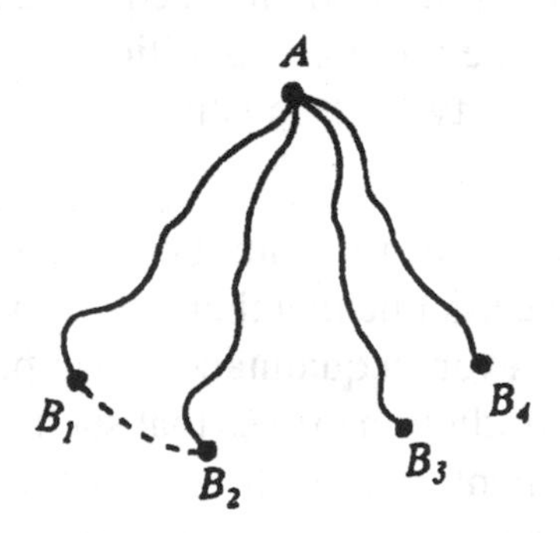

Fig. 9.5

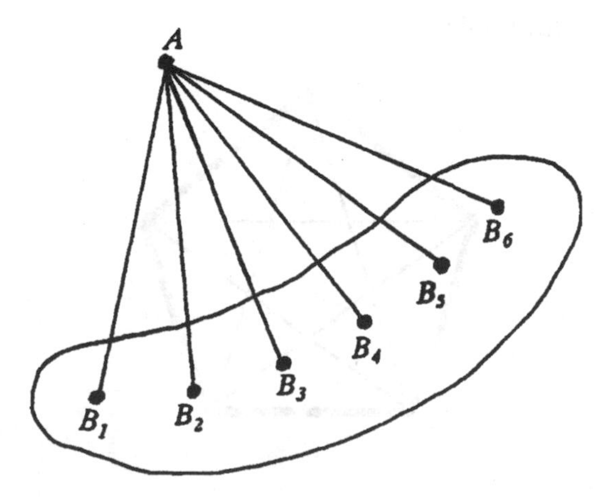

Fig. 9.6

mutual
acquaintances, then we are done. If three of them are mutual non-acquaintances, then three of them and A form four mutual non-acquaintances.

No doubt you noticed that the party-of-nine result is not symmetric: three acquaintances vs. four non-acquaintances, and there is nothing we can do about it. The good news is, however, that we can formulate a dual result that will also be true (do you see why?). *In any party of nine there are four mutual acquaintances or three mutual non-acquaintances.* ■

9.5. Behold (Figure 9.7):

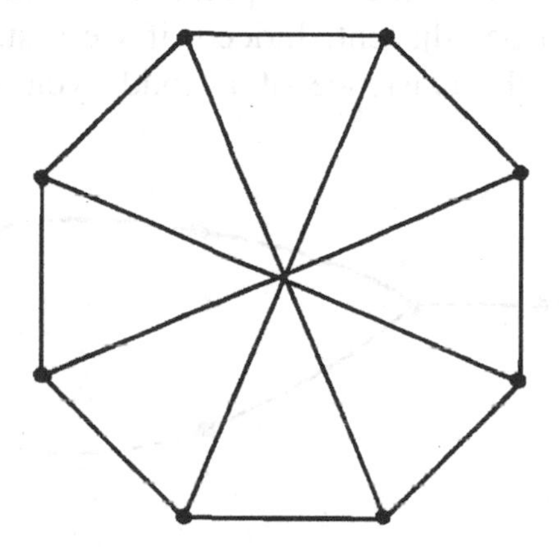

Fig. 9.7

9.6. As in the previous problems, it is convenient to represent people by vertices and connect "acquainted" vertices with edges. We need to partition the party into two subsets S_1 and S_2. We pick a vertex v and assign it to S_1. Now every vertex from which you can walk to v through a number of edges can be assigned the distance:

$$d(s, v) = \begin{cases} 0 \text{ if you walk through an even number of edges} \\ 1 \text{ if you walk through an odd number of edges} \end{cases}$$

This distance is uniquely defined because if we assume that one even and one odd walk exists from any w to v (Figure 9.8), then it is easy to show (do) the existence of an odd cycle of acquaintances.

Fig. 9.8

Now we assign the points w of distance 0 from v to the set S_1, and the points of distance 1 from v to the set S_2.

If any point v_t is not assigned, we assign v_t and all points of distance 0 from v_t to S_1, and the points of distance 1 from v_1 to S_2, and so on. (we have finitely many points).

Now we have to prove that two points x, y from the same subset, say S_1, cannot be adjacent. Indeed, if we assume the opposite, we can show (do) the existence of an odd cycle of acquaintances (Figure 9.9). ■

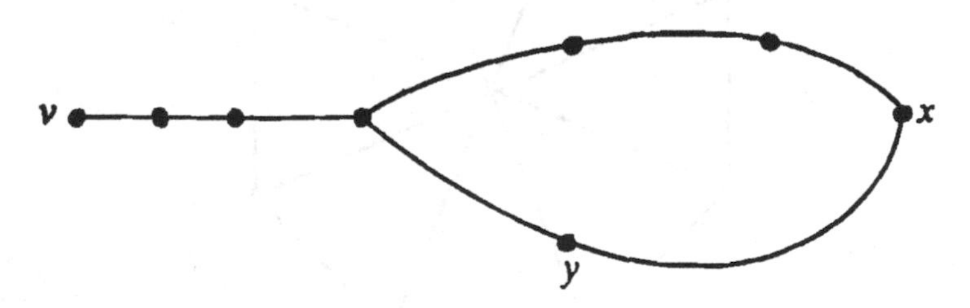

Fig. 9.9

9.7. Once again we represent people by vertices and adjacency between them by edges, and assume that the set of vertices is partitioned into two nonempty subsets S_1 and S_2 of mutual non-acquaintances (Figure 9.10).

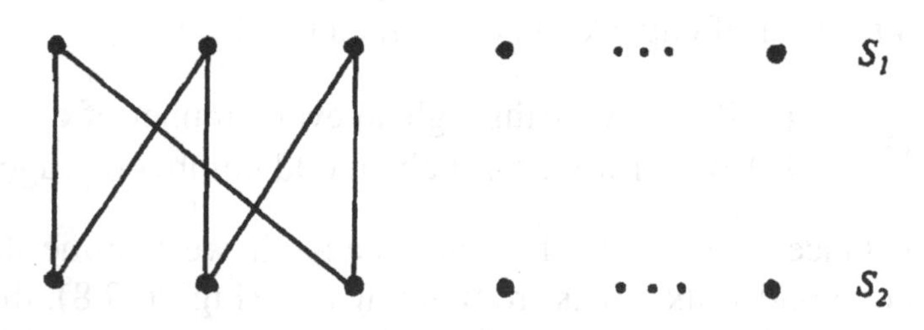

Fig. 9.10

The points in a cycle must alternate, one from S_1, one from S_2, one from S_1, again from S_2, and so on. This shows that any cycle of acquaintances must be even. ■

9.8. Assume that such a graph Gvexists. Then two distinct decompositions of n (n is the number of vertices of G) guarantee that when we present G as K_{n_1,n_2} and K_{m_1,m_2}, the corresponding decompositions $V = V_1 \cup V_2$ and $V = W_1 \cup W_2$ of the vertex set V of G are distinct. In other words, there are two vertices x and y in V such that both x and y belong to, say, V_1, and at the same time x is an element of W_1 and y is an element of W_2. But this means that x and y are not adjacent, and at the same time x and y are adjacent. This contradiction proves that such a graph G does not exist. ■

9.9. Every graph G can be viewed as the complement of its complement $\overline{G}$. ■

9.10. Let G be a graph with n vertices. Then, if any two vertices of G are adjacent, G contains a subgraph K_2. If no two vertices of G are adjacent, then $\overline{G} = K_n$. Therefore $R(2, n) = n$. ■

10 More About Graphs

Graphs appear in our discussion as diagrams of acquaintances. Thus, the only thing that matters when we represent a graph in the plane is the set of vertices (but not their positions on the plane) and which vertices are adjacent (but not the shapes of the edges, which we presume have no points in common except the vertices of the graph). In fact, think of a graph as a set of pins, some of which are connected by rubber bands. A graph remains the same if we reposition the pins and stretch the rubber bands.

Two graphs are called *isomorphic* if "pins" of one of them can be repositioned and its "rubber bands" stretched so that the two become identical.

More formally, two graphs G and G_1 are said to be *isomorphic* if there is a one-to-one correspondence $f\colon V \to V_1$ of their vertex sets that preserves adjacency, i.e., vertices v_1 and v_2 of G are adjacent if and only if $f(v_1)$ and $f(v_2)$ of G_1 are adjacent.

We denote the isomorphism of graphs G and G_1 by $G \cong G_1$.

Example 10.1. Are the graphs in Figure 10.1 isomorphic?

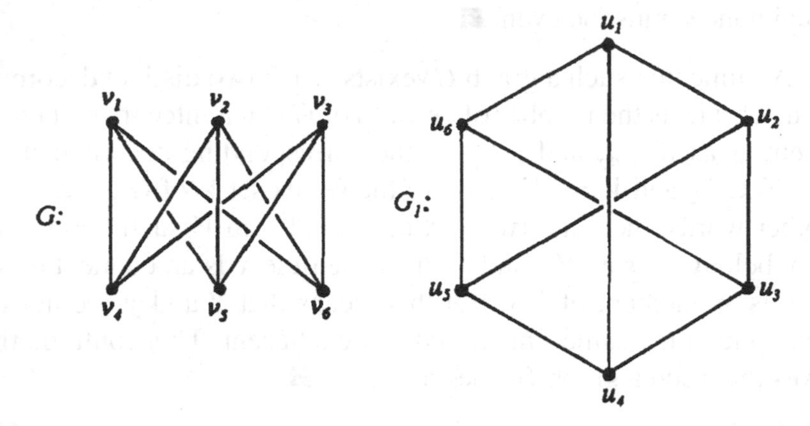

Fig. 10.1 Isomorphic graphs

Solution. Let us manipulate "pins" and "rubber bands" of G : first we flip v_2v_5, then stretch it (Figure 10.2).

Lo and behold, we end up with a graph identical to G_1. G and G_1 are isomorphic. ■

Needless to say, two isomorphic graphs *must* have equal numbers of vertices and edges; therefore, the numbers of vertices and edges are what we call *invariants* of graphs (invariants are characteristics shared by all isomorphic graphs). The equality of these two invariants, however, is not sufficient to prove the isomorphism of two graphs.

Example 10.2. Are the graphs in Figure 10.3 isomorphic?

Solution. Even though G and G_1 have equal numbers of vertices and edges, they are not isomorphic. Under no one-to-one correspondence of vertices can adjacency be preserved, because G_1 contains a vertex v of degree one. This vertex must correspond under isomorphic correspondence to a vertex of degree one in G. But G has no such vertex! ■

While solving Example 10.2, we discovered a new invariant of graphs: the degrees of its vertices. Do we have enough invariants to guarantee isomorphism of two graphs? In other words, given G and G_1 with equal numbers of vertices and edges and equal sequences of vertex degrees, do G and G_1 have to be isomorphic?

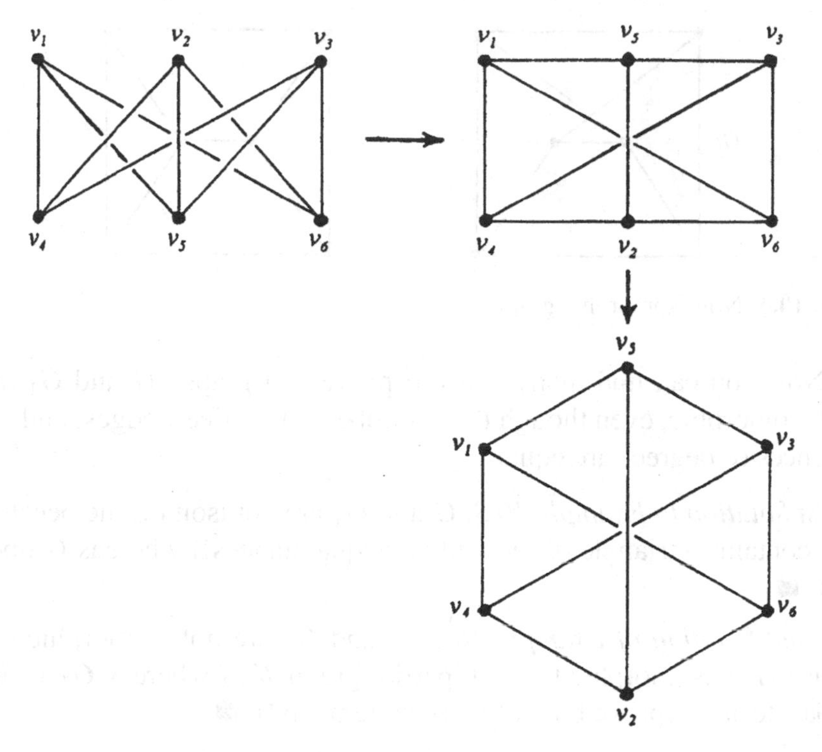

Fig. 10.2

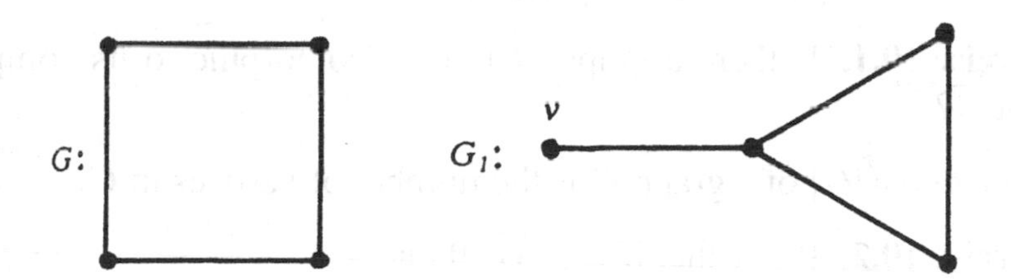

Fig. 10.3 Non-isomorphic graphs

Example 10.3. Are the graphs in Figure 10.4 isomorphic?

You may have sensed that any characteristic of G that can be expressed in terms of adjacency will be an invariant (i.e., shared by all graphs isomorphic to G). Such invariants include length of cycles, complete subgraphs, etc.

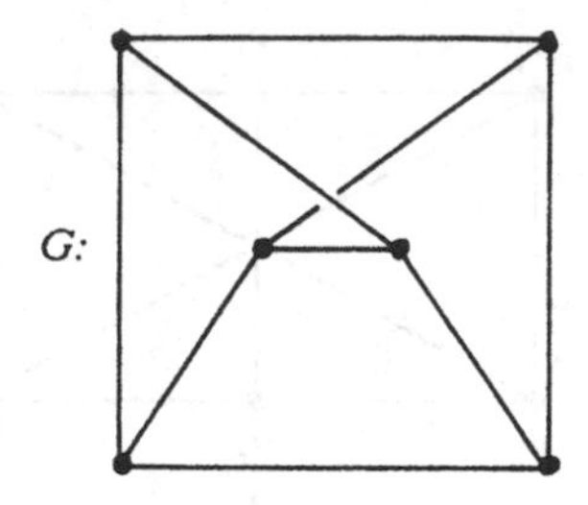

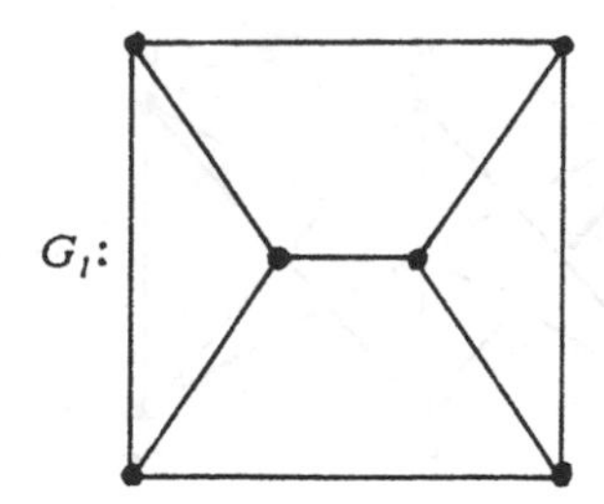

Fig. 10.4 Non-isomorphic graphs

Now you can find many ways to prove that graphs G and G_1 are not isomorphic, even though their numbers of vertices, edges, and sequences of degrees are equal.

First Solution to Example 10.3. G and G_1 are not isomorphic because G_1 contains a triangle (three mutual acquaintances!) whereas G does not. ■

Second Solution of Example 10.3. G and G_1 are not isomorphic because G is isomorphic to the bipartite graph $K_{3,3}$ whereas G_1 is not bipartite at all (prove both of these statements!). ■

As we were writing these lines for you, the following problem came to mind: *find all positive integers n such that there is a graph of order n that is isomorphic to its complement.*

Exercise 10.1. Is there a graph G that is isomorphic to its complement $\overline{G}$?

The *order* $|G|$ of a graph G is the number of vertices in G.

Exercise 10.2. Prove that if $G \cong \overline{G}$, then

$$|G| \equiv 0 \pmod 4$$

or

$$|G| \equiv 1 \pmod 4.$$

Exercise 10.3. Is there a graph of order 4 that is isomorphic to its complement?

Exercise 10.4. Is there a graph of order 8 that is isomorphic to its complement?

Now we are ready to attack the following theorem.

Theorem 10.1. *There is a graph of order n that is isomorphic to its complement if and only if*

$$n \equiv 0 \pmod 4 \quad \text{or} \quad n \equiv 1 \pmod 4. \tag{$*$}$$

Proof. The necessity of condition ($*$) was proven in Exercise 10.2 above.

Let us prove the sufficiency.

For $n = 4$, the graph G_4 and its complement $\overline{G}_4$ are presented in Figure 10.5.

To construct G_8 we take two solutions for $n = 4$ and *sew* them together with a cycle of length 8 (Figure 10.6). The complement $\overline{G}_8$ (Figure 10.7) is graphically identical to G_8.

Now we generalize this construction to create G_{4m} for any positive integer m.

In order to construct the complete graph K_m, we start with m vertices and connect every two of them together. Similarly, here we start with m copies of the graph G_4 (Figure 10.5) and *sew* every two of them together with a cycle of length 8. We get the graph G_{4m} that is

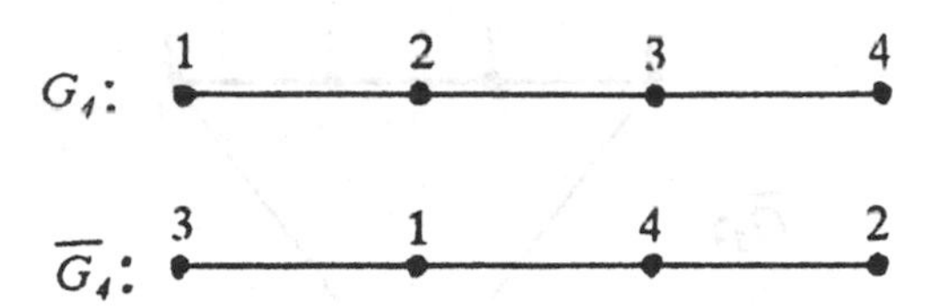

Fig. 10.5

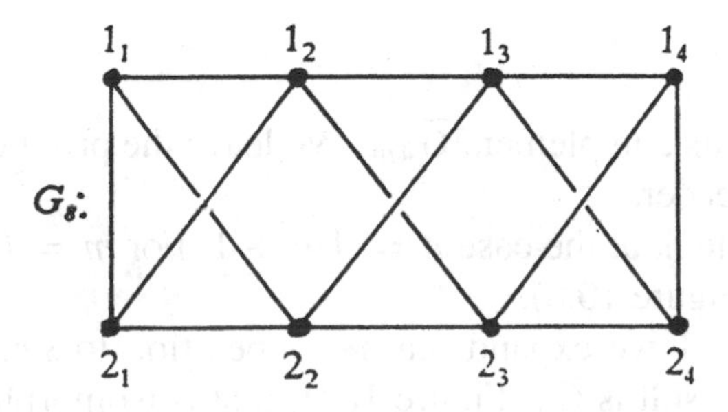

Fig. 10.6

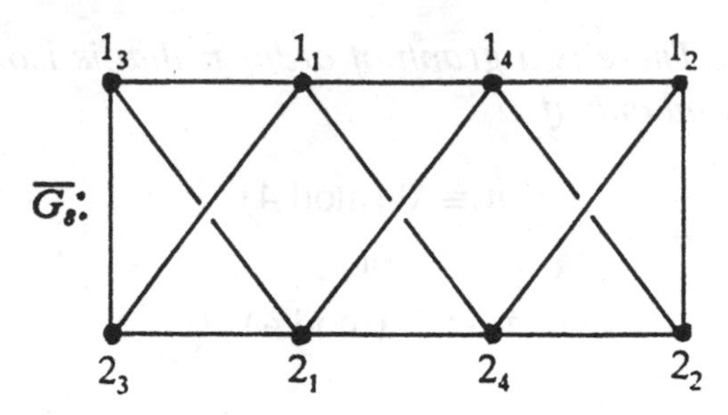

Fig. 10.7

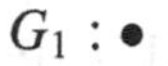

$G_1 : \bullet$

Fig. 10.8

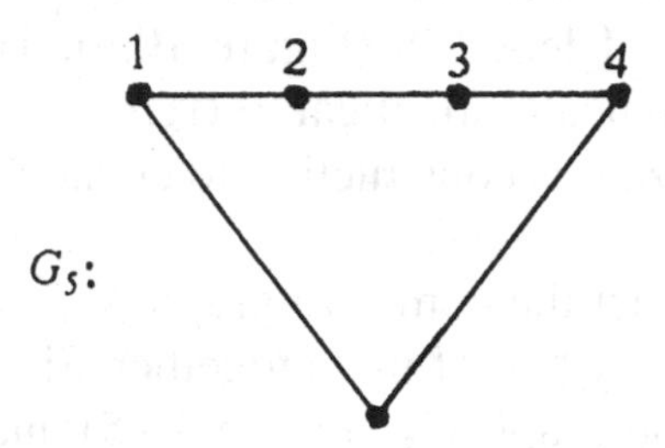

Fig. 10.9

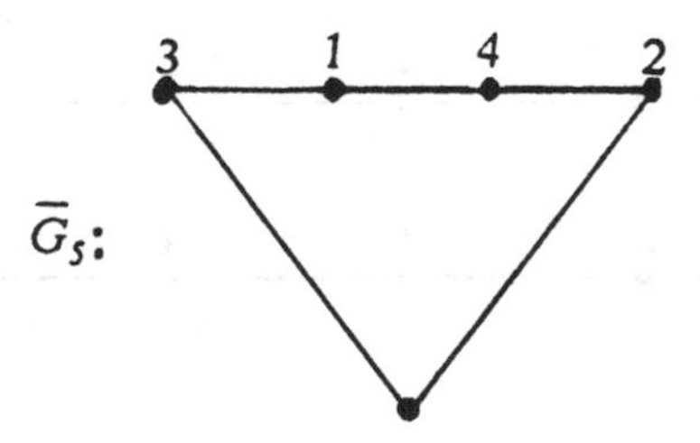

Fig. 10.10

isomorphic to its complement $\overline{G}_{4m}$. We leave the proof of the isomorphism to the reader.

Now let us look at the case $n = 4m + 1$. For $m = 0$, the solution G_1 is trivial (Figure 10.8).

When $m = 1$ we expand our *sew* operation to *sew* G_1 and G_4 together. The result is G_5 (Figure 10.9) that is isomorphic to its complement $\overline{G}_5$ (Figure 10.10).

Now you can guess how we construct G_{4m+1} for any positive integer m. We start with m copies of G_4 and one copy of G_1 and *sew* every two of these $m + 1$ basic pieces together. We get G_{4m+1} that is isomorphic to its complement $\overline{G}_{4m+1}$. We would like our readers to prove this isomorphism on their own. ■

Exercise 10.5. Is there a graph G isomorphic to its complement $\overline{G}$ and not isomorphic to any of the graphs G_n constructed in the proof of Theorem 10.1?

Let G be a graph of order n.G and $\overline{G}$ first appeared when we colored the edges of the complete graph K_n in two colors.

In Theorem 10.1 we found G_n such that $G_n = \overline{G}_n$. In a sense, we *tiled* the complete graph K_n with two copies of G_n.

Let k be a positive integer. We can color the edges of K_n in k colors and pose a question: *find all n such that K_n can be tiled with k copies of a tile G_n.*

Exercise 10.6. Prove that if K_n is tileable by m copies of G_n, then

$$m \text{ is a divisor of } \binom{n}{2}, \qquad (*)$$

where $\binom{n}{2} = \dfrac{n(n-1)}{2}$.

It is relatively easy to show that *the condition $(*)$ is also sufficient for any odd k.* For example, let us prove it for $m = 5$ and call this **Example 10.4**.

Solution to Example 10.4. Due to Exercise 10.6, we must have

$$n \equiv 0 \pmod 5$$

or

$$n \equiv 1 \pmod 5.$$

For $n = 1$, the result is obviously true. We can easily solve for $n = 5$: in Figure 10.11, bold lines show the edges of one of the five tiles G_5 (other tiles can be obtained by rotations of G_5 about the center of the pentagon).

Now, similarly to the proof of Theorem 10.1, we need to show how to *sew up* solutions for larger values of n.

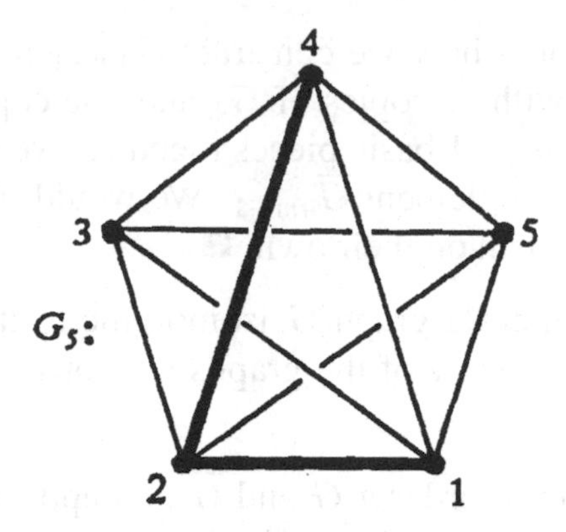

Fig. 10.11

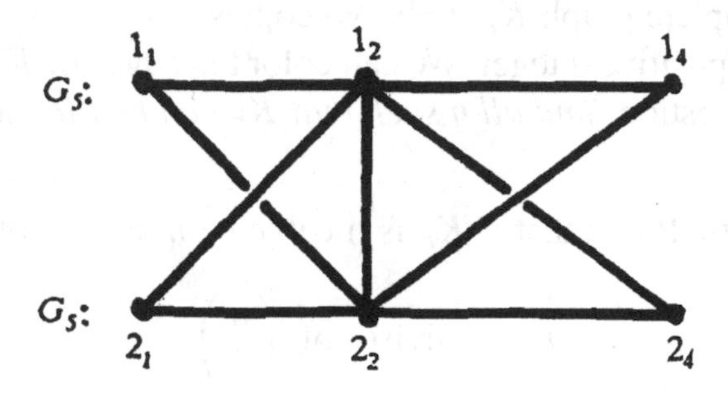

Fig. 10.12

G_1 : •

Fig. 10.13

For $n = 10$ we take two copies of G_5 and *sew* them together to get G_{10}, as shown in Figure 10.12. We do this, of course, for each of the five colors.

Let $n = 5m$, where m is an integer greater than 1. We construct m copies of G_5 and *sew* every two of them together as shown in Figure 10.12 to get G_{5m}. We do this for each of the five colors. As a result, K_{5m} is tiled with five copies of the tile G_{5m}.

Now let us solve the case $n = 5m + 1$.

For $m = 0$ the solution G_1 is trivial (Figure 10.13). When $m = 1$ we expand our *sew* operation to connect G_1 and G_5. The result is G_6 shown in Figure 10.14.

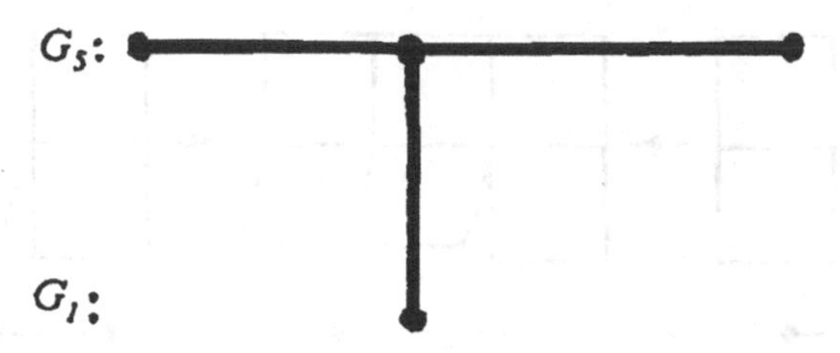

Fig. 10.14

G_{5m+1} ($k \geq 1$) is constructed from m copies of G_5 and one copy of G_1 by *sewing* every two of them together. It is not difficult to show (do) that K_{5m+1} can be tiled with five copies of G_{5m+1}. ■

This construction allows a straightforward generalization.

Exercise 10.7. For every odd divisor m of $\binom{n}{2}$, construct a tile graph G_n such that K_n can be tiled by m copies of G_n.

This problem has been solved completely by Frank Harary, R.W. Robinson, and Nicholas Wormald in [HRW]. They showed that the condition (∗) is always sufficient.

In Chapter 1 we discussed tiling of rectangular boards by polyominal tiles. This problem can be translated into the language of graph theory. Indeed, we can represent every unit square of the board by a vertex and connect two vertices with an edge if and only if the corresponding squares share a side (Figure 10.15).

And, of course, a polyominal tile will become a connected graph.

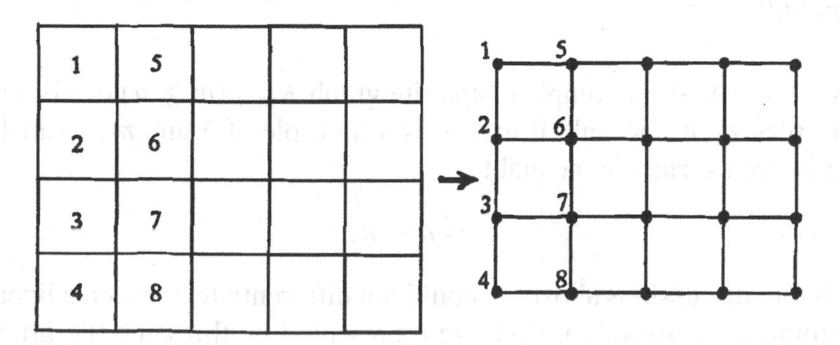

Fig. 10.15

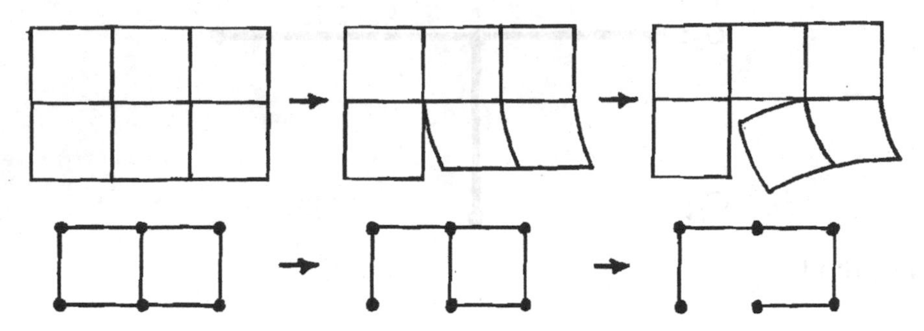

Fig. 10.16

Tiling a rectangle is essentially no different from cutting it into tiles. But cutting the board along the checkered lines translates into cutting edges of the graph (Figure 10.16). For simplicity, *cutting a graph G* equates to cutting (removing) some of the edges of G. Now we can translate the whole problem.

Given two connected graphs G and T. Cut G into components each isomorphic to T or simply into tiles T.

Exercise 10.8. If a graph G can be cut into components T, then |T| is a divisor of $|G|$.

Exercise 10.9. Let P_3 be a tromino graph (Figure 10.17). Then K_n can be cut into tiles P_3 if and only if n is a multiple of 3.

Fig. 10.17

Exercise 10.10. A complete bipartite graph $K_{m,n}$ $(m \geq n)$ can be cut into tiles P_3 if and only if $m + n$ is a multiple of 3 and m, n satisfy the isosceles triangle inequality:

$$m \leq n + n.$$

In our discussions above we could not differentiate between a linear tromino and L-tromino. Both are represented by the same tile graph P_3 of Figure 10.17. We can resolve this situation by coloring edges of a graph in two colors.

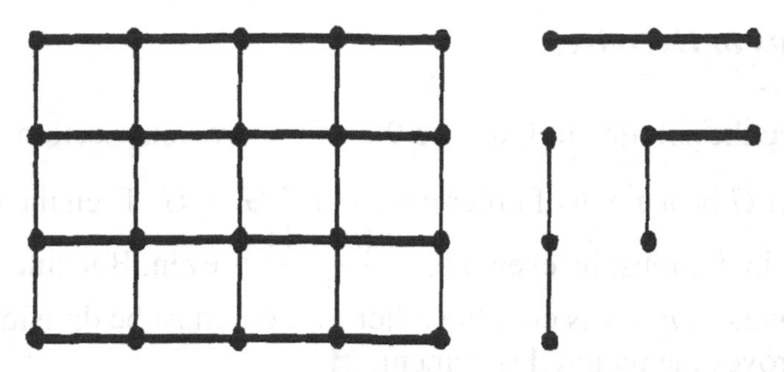

Fig. 10.18

Let G be a graph of a rectangular board (Figure 10.18). Color all horizontal edges in one color (bold segments in Figure 10.18) and all vertical edges in the other color. Now we can differentiate between the two types of trominoes. The linear tromino produces a tile P_3 with both edges of the same color, while L-tromino corresponds to a file P_3 with one edge of each color.

Thus, the problem of tiling with L-trominos can be translated into the language of graphs as follows:

The edges of a connected graph G are colored in two colors. Cut G into tiles P_3 containing one edge of each color.

The problem of packing a parallelepiped can also be translated into the language of graph theory. We represent each unit cube of the parallelepiped by a vertex and connect two vertices by an edge if and only if the corresponding unit cubes share a face (Figure 10.19).

If we wish to differentiate between various 3-dimensional tiles, we may color the edges of the graph in three colors in accordance with the coordinate axes.

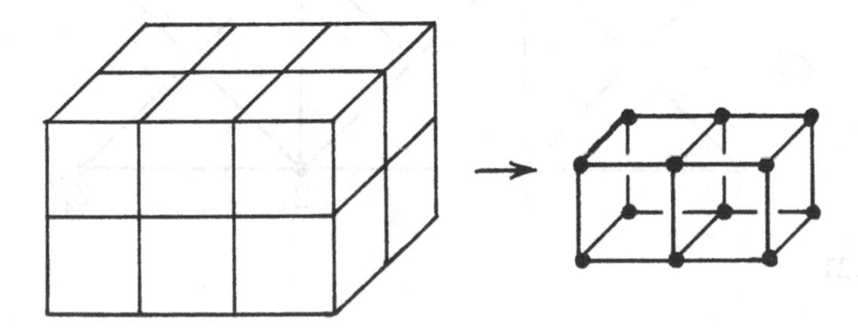

Fig. 10.19

Solutions to Exercises

10.1. See the solution to Exercise 9.3 of the previous section. ■

10.2. Let G be a graph of order n such that $G = \overline{G}$. Then the number of edges in K_n must be even, i.e., $\frac{n(n-1)}{2}$ is even. But since one of the numbers n, $n-1$ is odd, the other of them must be divisible by 4, which proves the required statement. ■

10.3. Behold (Figure 10.20):

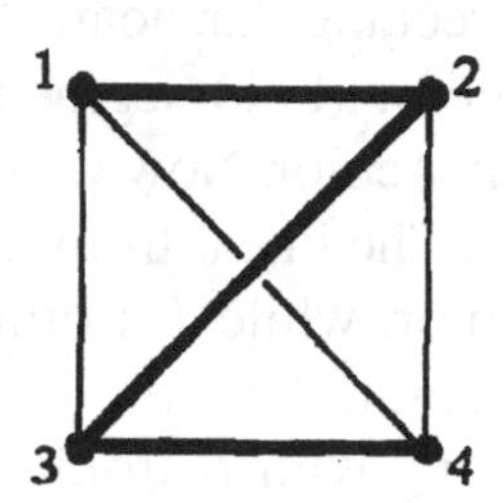

Fig. 10.20

10.4. See Figure 10.6. ■

10.5. Yes, in fact, for any $m > 1$ we can construct an alternative $G'_{4m} \neq G_{4m}$, and for any $m \geq 1$ an alternative $G'_{4m+1} \neq G_{4m+1}$. We can *sew* the basic pieces differently, see Figures 10.21 and 10.22. ■

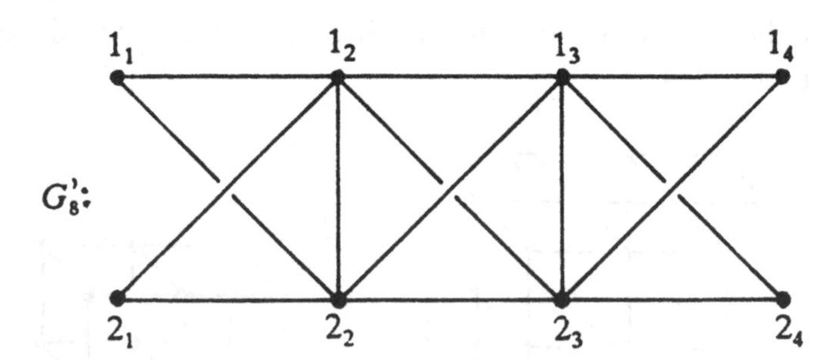

Fig. 10.21

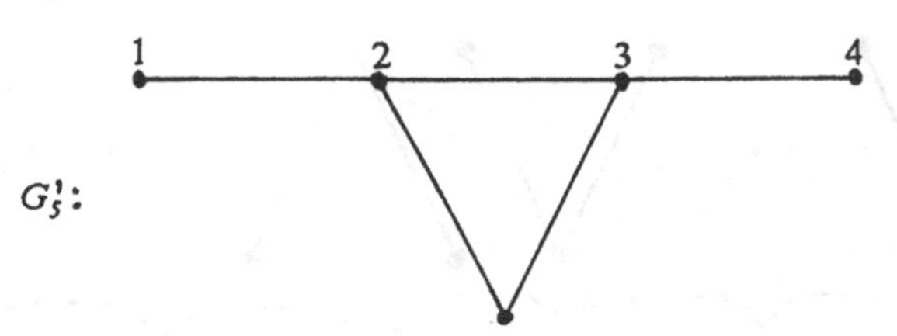

Fig. 10.22

10.6. Obviously, m must be a divisor of the number of edges of K_n, which is exactly $\binom{n}{2}$. ■

10.7. The construction is similar to the one in Example 10.4. ■

10.8. When the graph G is partitioned into tiles T, $|G|$ vertices are partitioned into groups of $|T|$ vertices, therefore $|T|$ is a divisor of $|G|$. ■

10.9. Due to Exercise 10.8, n must be a multiple of 3, i.e., $n = 3k$. If $n = 3k$, we can partition n vertices into groups of three, save in every group two edges, and cut all other edges. ■

10.10. Let V_1 and V_2 be subsets of m and n nonadjacent vertices, respectively, that form the graph $K_{m,n}$ ($m \geq n$). Assume that $K_{m,n}$ is cut into tiles P_3. Then at least one vertex of each tile must belong to V_2 (remember, there are no edges between any two vertices of V_1). Therefore, $m \leq n + n$. In addition, $m + n$ is a multiple of 3 (Exercise 10.8).

Conversely, assume that positive integers m, n satisfy the inequality n $\leq m \leq n + n$ and $m + n$ is a multiple of 3. Let us take a look at $2m - n$ and $2n - m$. They are positive. Both are divisible by 3 because $2m - n = 3m - (m + n)$ and $2n - m = 3n - (m + n)$. Now the construction is clear. We line up $\frac{2m - n}{3}$ copies of "V" followed by $\frac{2n - m}{3}$ copies of "A" (Figure 10.23).

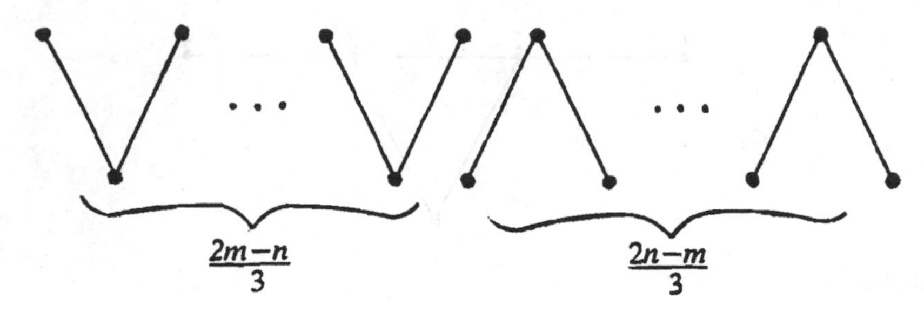

Fig. 10.23

In the top row of Figure 10.23 we have

$$2\frac{2m-n}{3}+\frac{2n-m}{3}=m$$

vertices. In the bottom row we have

$$2\frac{2n-m}{3}+\frac{2m-n}{3}=n$$

vertices. Therefore, Figure 10.23 shows exactly how to cut $K_{m,n}$ into P_3 tiles. ■

11 Planarity

How would you draw a complete graph K_4 in the plane? Probably as in Figure 11.1:

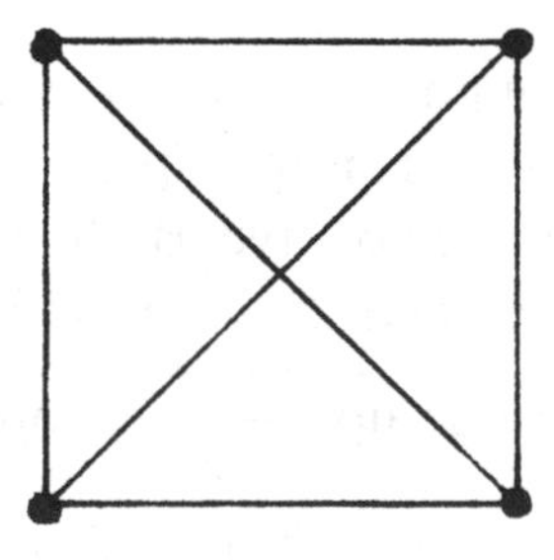

Fig. 11.1

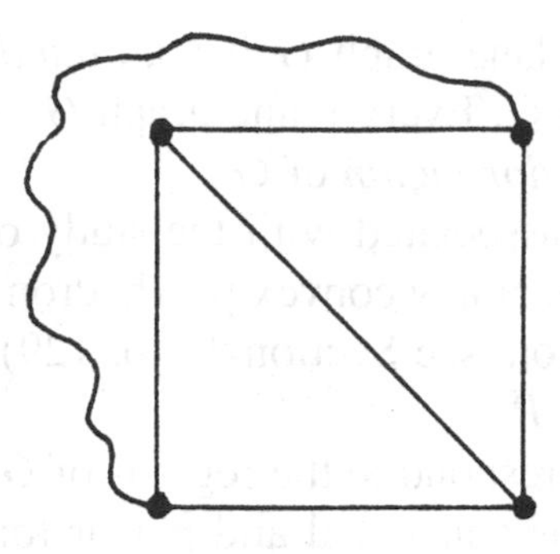

Fig. 11.2

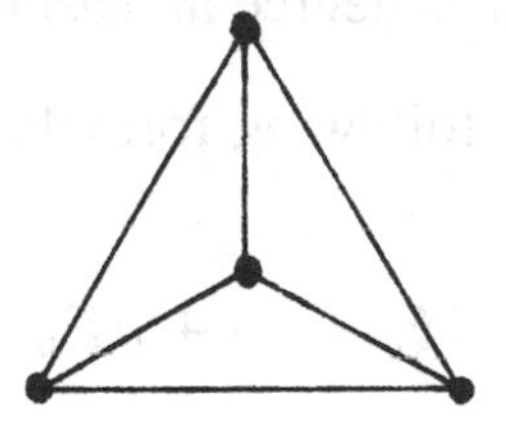

Fig. 11.3

In Figure 11.1, two edges of K_4 intersect. Can we avoid it, i.e., draw K_4 in the plane without any intersections of edges? Yes, we can.[3] Remember, we can think of a graph as a set of pins, some of which are connected by rubber bands. So we pick up one of the diagonal rubber bands and pull it out (Figure 11.2). We can also redraw K_4 as in Figure 11.3.

A graph that *can* be drawn in the plane without any intersection of edges is *planar*. Otherwise, a graph is *non-planar.* If a planar graph *is* already drawn in the plane without intersections of edges, it is called a *plane graph*.

A region of a plane graph G is the maximal part of the plane for which any two points may be joined by a curve that does not intersect edges or vertices of G. If we think of the plane as a sheet of paper, the regions of G are the pieces into which the plane is divided when we cut it along the edges of G.

[3] This sentence was made famous by president Barak Obama's 2008 campaign.

Each region of a plane graph G has a *boundary* that consists of vertices and edges of G. Every plane graph G also has a unbounded region called the *exterior region* of G.

Planar graphs are associated with the study of polyhedra. In fact, the vertices and edges of any convex polyhedron P (for the definition of a convex polyhedron, see Section 13, p. 129) form a graph $G(P)$ called the *skeleton* of P.

The faces of P correspond to the regions of $G(P)$.

This graph $G(P)$ is connected and planar for every convex polyhedron P (can you prove that?). Therefore, everything we prove for planar graphs will be applicable to the skeletons of polyhedra.

Every vertex of $G(P)$ has degree at least three.

Exercise 11.1. Prove the following formula for the graph $G(P)$ of a polyhedron P:

$$3F_3 + 4F_4 + \cdots + nF_n = 2q, \tag{1}$$

where q is the number of edges of P, F_3 is the number of triangular faces of P, F_4 is the number of quadrangular faces, ..., F_n is the number of n-gonal faces (where n is the greatest number of edges bounding a face of P).

Exercise 11.1 has the following very important corollary. Can you prove it?

Corollary 11.1. *For any planar graph G,*

$$3F_3 + 4F_4 + \cdots + nF_n \leq 2q, \tag{1'}$$

where q is the number of edges of G and $F_3, F_4, \ldots, F_n$ are the numbers of triangular, quadrangular, ..., n-gonal faces of G, respectively.

Let us prove the celebrated formula discovered by one of the greatest mathematicians of all time, Leonard Euler (1707–1783).

Theorem 11.1. *(Euler's Formula for Graphs) If G is a connected plane graph with p vertices, q edges, and r regions, then*

$$p + r = q + 2. \tag{2}$$

Proof. We will prove (2) by induction on q.

For $q = 1$ the result is obvious since in this case $p = 2$ and $r = 1$.

Assume (2) is true for all connected plane graphs with fewer than q edges.

Let G be a connected plane graph with q edges (and p vertices and r regions). We will consider two cases.

Case 1. If G has no cycles, then G contains a vertex of degree one. Indeed, of G we can travel along an edge to an adjacent vertex v_2 from any vertex v_1. If v_2 is not of degree one, we continue our travel to v_3, etc. Since we can never come again to any of the vertices we previously visited (there are no cycles, remember?), and the total number of vertices is finite, we must end at a vertex v_n of degree one. Now we remove v_n together with the edge incident to v_n. We get a graph G' with $q-1$ edges, $p-1$ vertices and r regions. Since by the inductive assumption (2) is true for G', we get

$$(p-1)+r=(q-1)+2.$$

By adding 1 to both sides of the last equality, we prove the required equality (2) for the graph G.

Case 2. If G has a cycle, we remove one edge of the cycle. We get a plane graph G' that is connected (do you see why?) and has $q-1$ edges, p vertices, and $r-1$ regions. By the inductive assumption, the equality (2) is true for G', since

$$p+(r-1)=(q-1)+2$$

implies (2). ■

As we mentioned before, the results from the theory of planar graphs can be applied to polyhedra. Thus we get the following important and beautiful corollary.

Corollary 11.2. *(Euler's Formula for Polyhedra) If V, E, and F are the numbers of vertices, edges, and faces of a convex polyhedron, then*

$$V+F=E+2. \tag{3}$$

Exercise 11.2. Prove that for any planar graph with p vertices and q edges,

$$q \leq 3(p-2). \tag{4}$$

Exercise 11.3. Prove that for any planar graph with p vertices and q edges and containing *no* triangles,

$$q \leq 2(p-2). \tag{5}$$

Exercise 11.4. If a graph G contains a non-planar subgraph, then G is non-planar.

Exercise 11.5. Prove that a complete graph K_n is planar if and only if $n < 5$.

Exercise 11.6. Prove that a complete bipartite $K_{m,n}$ is planar if and only if at least one of the positive integers m, n is less than three.

An *elementary subdivision* of a graph G is a graph (with at least one edge) obtained by inserting an additional vertex at an inside point of some edge of G. Two graphs G_1 and G_2 are said to be *homeomorphic* if each of them "comes from the same ancestor" G, i.e., both can be obtained from the same graph G by a succession of elementary subdivisions.

Exercise 11.7. Prove that the graphs G_1 and G_2 in Figure 11.4 are homeomorphic.

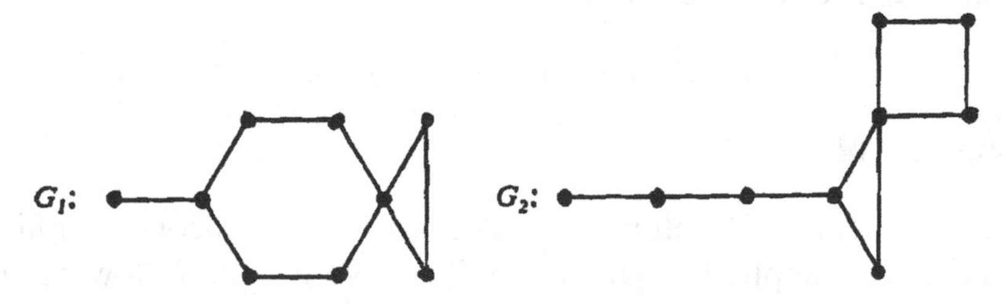

Fig. 11.4 Homeomorphic graphs

Now we are ready for the brilliant criterion of planarity discovered by the famous Polish mathematician Kazimierz Kuratowski (1896–1980) in 1930 ([Ku]).

Kuratowski's Theorem 11.2. A graph is planar if and only if it contains no subgraph homeomorphic to K_5 or $K_{3,3}$.

Exercise 11.8. Prove that the Petersen graph in Figure 11.5 is non-planar.

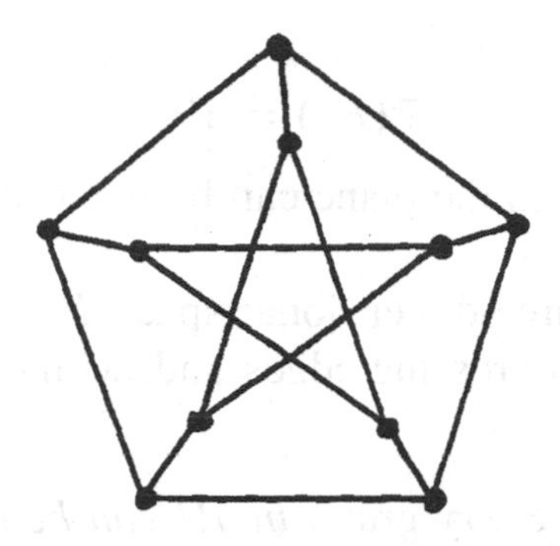

Fig. 11.5 The Petersen Graph

Exercise 11.9. Any graph can be drawn in the plane so that every edge is a straight line segment.

The following result, though, is not at all obvious.

Theorem 11.3. *(Wagner [W], Fáry [F]) A planar graph can be drawn in the plane without intersection of edges, and all edges being straight line segments.*

We have divided all connected graphs into planar and non-planar. Are all non-planar graphs alike? No, and we have a couple of invariants for you.

There are many ways to draw a graph in the plane, and, of course, the number of crossings of edges may vary. The *crossing number* ν(G) of a graph G is the minimum number of crossings of edges among all the drawings of G in the plane.

The *rectilinear crossing number* $\overline{\nu}(G)$ of G is the minimum number of crossings of edges taken over all those drawings of G in the plane in which every edge is a straight line segment.

Of course,

$$\nu(G) \leq \overline{\nu}(G).$$

But do we really need two crossing numbers? For planar graphs $\nu(G) = \overline{\nu}(G) = 0$ (Theorem 11.3). Moreover, Richard Guy proved [Gu] that for any graph G with seven points or less $\nu(G) = \overline{\nu}(G)$.

Yet, in the same paper he showed that the complete graph K_8 delivers a counterexample:

$$\nu(K_8) = 18$$

but

$$\overline{\nu}(K_8) = 19.$$

So, not everything in the plane can be reduced to drawing straight line segments.

What about the three-dimensional space R^3? Can every graph in R^3 be drawn without crossing edges and, with every edge a straight line segment?

Theorem 11.4. *Yes! Every graph in R^3 can be drawn so that every edge is a straight line segment and no edge crossing occurs.*

Proof. Place the vertices of the given graph G in the *points of general position*, i.e., so that no four points are in a plane, and no three points are on a line. Now connect the adjacent points of G by line segments. No crossing will occur, because otherwise four vertices would lie in a plane.

We are done. Are we?

Does a set of n points of general position exist for any positive integer n? Yes; we can construct it by induction.

The first two distinct points we can pick arbitrarily. We choose the third point so that not all three are on a line.

Assume that n points of general position are chosen ($n \geq 3$). Through every three points we draw a plane. We get finitely many planes (namely $m = \binom{n}{3}$ planes). They do not exhaust the space so we can pick the $(n+1)$st point outside of the drawn planes.

We are done. Are we?

Well, yes, if you believe that m planes (m being an integer) do not exhaust the space. If you do not believe it, pick a plane P not parallel to (and not coincidental with) any of the m planes. The m planes intersect P along m straight lines, which do not exhaust P, so we can pick the $(n+1)$st point on P outside of the m lines.

We are done. Are we?

Well, yes, if you believe that m straight lines do not exhaust a plane P. If you do not believe it, pick a straight line L in P that is not parallel to (or coincidental with) any of the m lines. The m lines intersect L at

m points, which do not exhaust L, so we can pick the $(n+1)$st point in L, distinct from the m points.

We are done. Are we? ■

Solution to Exercises

11.1. There are two ways to add up the T edges on the boundaries of all faces. Triangular faces are bounded by three edges, quadrangular by four, ..., n-gular faces by n edges. Therefore,

$$T = 3F_3 + 4F_4 + \cdots + nF_n.$$

On the other hand, T counts every edge exactly twice because every edge e is counted on the boundary of the two faces that are separated by e. Therefore, $T = 2q$. ■

11.2. The inequality (1′) of Corollary 11.1 implies

$$3(F_3 + F_4 + \cdots + F_n) \le 2q,$$

i.e.,

$$3r \le 2q, \tag{6}$$

where r is the total number of regions of G.

Combining the Euler formula (2) with (5), we get

$$\left(p + \frac{2}{3}q\right) \ge q + 2,$$

i.e.,

$$q \le 3(p-2). \blacksquare$$

11.3. Since $F_3 = 0$, the inequality (1′) of Corollary 11.1 implies

$$4(F_4 + \cdots + F_n) \le 2q,$$

i.e.,

$$4r \le 2q, \tag{7}$$

where r is the total number of regions of G. Now the Euler formula (2) and (6) combine to produce the required inequality:

$$p + \frac{1}{2}q \ge q + 2,$$

i.e.,

$$q \leq 2(p-2). \blacksquare$$

11.4. Argue by contradiction. ■

11.5. We already know that K_n is planar for $n < 5$ (Figure 11.1 shows this for $n = 4$). Let us take a look at K_5. It has $p = 5$ vertices and $q = 10$ edges. Assume that K_5 is planar; then, by Exercise 11.2, it satisfies the inequality $q \leq 3(p-2)$, i.e., $10 \leq 9$, which is absurd. Therefore, K_5 is not planar.

Finally, every graph K_n for $n \geq 5$ contains K_5 as a subgraph. Therefore, K_n is not planar for $n \geq 5$. ■

11.6. Figure 11.6 shows that $K_{m,2}$ is planar for any positive integer m. Therefore any subgraph of $K_{m,2}$ is planar as well, i.e., a complete bipartite graph $K_{m,n}$ is planar if at least one of the positive integers m, n is less than three.

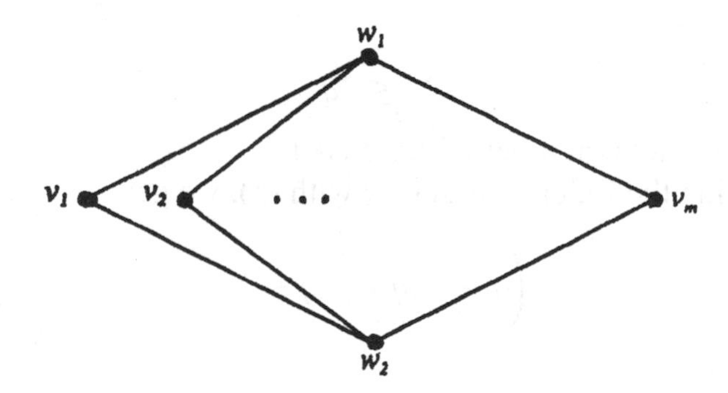

Fig. 11.6

Let us take a look at $K_{3,3}$. It has $p = 6$ vertices, $q = 9$ edges, and no triangles (Theorem 9.2). Therefore, by Exercise 11.3′, $q \leq 2(p-2)$, i.e., $9 \leq 8$, which is absurd. Therefore, $K_{3,3}$ is not planar.

Every graph $K_{m,n}$ for $m \geq 3, n \geq 3$ contains $K_{3,3}$ as a subgraph, and therefore is not planar. ■

11.7. Both graphs G_1 and G_2 can be obtained from the graph G (Figure 11.7) by a succession of elementary subdivisions. ■

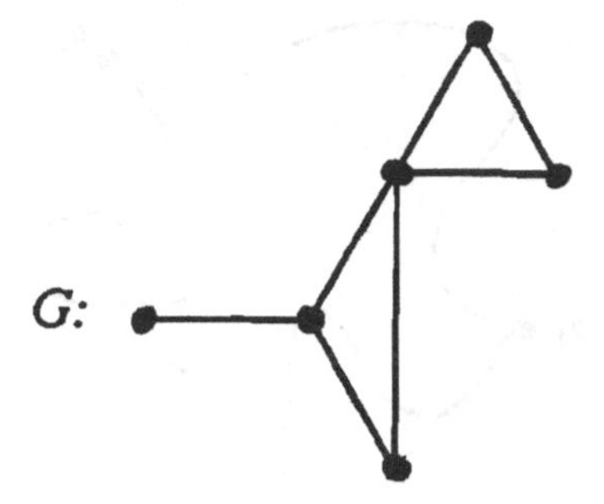

Fig. 11.7

11.8. Behold (Figure 11.8):

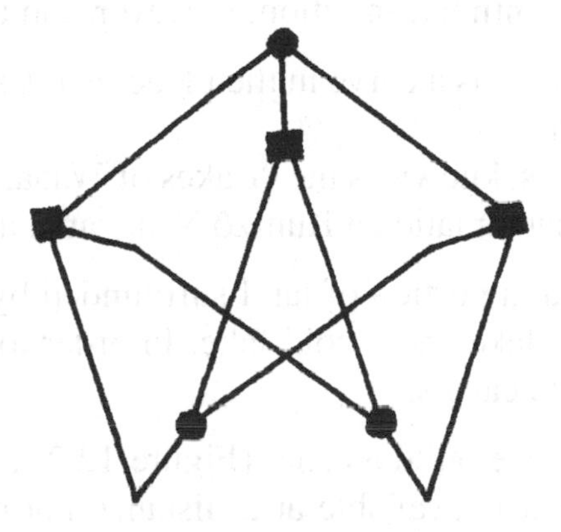

Fig. 11.8

12 The Intersection Index and the Jordan Curve Theorem

In the previous section we talked about the *regions* we obtain when a graph is drawn in the plane without intersections. Is it clear what a *region* is? Perhaps you would answer *yes.* But allow us to shake your confidence in the clarity of this notion.

Let us consider a very simple example: look at a complete graph K_3, the contour of a triangle (Figure 12.1). Is it evident that the graph in Figure 12.1 divides the plane into two regions? In other words, is it true that every closed curve that does not intersect itself defines two regions, interior and exterior?

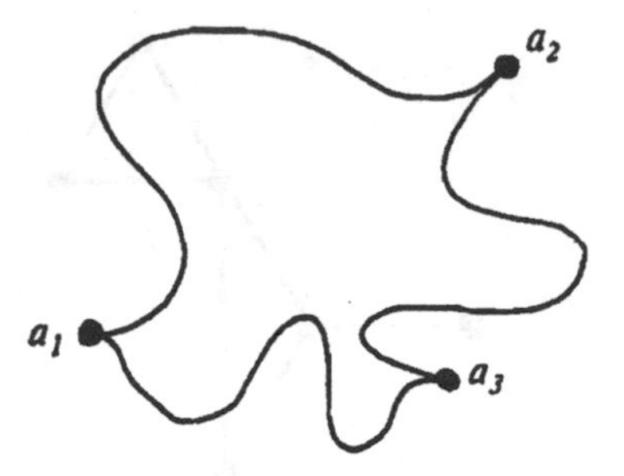

Fig. 12.1

It appears "visually obvious" that a curve C cannot be a boundary common to more than two regions in the plane such that all regions border C along its entirety. Intuition, however, can trick us.

Example 12.1. There is a curve in the plane that is a common boundary of three regions.

Some such curves, known as the "Lakes of Wada," were discovered by the Japanese mathematician Kunizô Yoneyama in 1917 ([Y]).

Proof. Assume that a portion of land surrounded by the sea contains two lakes: a warm lake and a cold lake. In order to provide the land with water we build canals.

On the first day, we build a canal (Figure 12.2) that delivers warm lake water so that it is available at a distance not exceeding 1 from

Fig. 12.2 Lakes of Wada

every point of land (this canal is neither connected to the sea nor to the cold lake!).

On the second day, we build a canal that delivers cold lake water so that it is available at a distance not exceeding 1 from every point of the remaining land (this canal is not connected to the sea, the warm lake, or to the previously built canal).

On the third day, we build a canal that delivers sea water so that it is available at a distance not exceeding 1 from every point of the remaining land (of course, this canal is not connected to the lakes or to the previously built canals).

During the next three days we extend the three canals further so that the warm lake water, the cold lake water, and the sea water are available from any point of the remaining land at a distance not exceeding $\frac{1}{2}$.

During the following three days we increase the density of the canals so that every type of water is available from every point of the remaining land at a distance not exceeding $\frac{1}{4}$, etc. Note that after each day, the remaining land is a connected piece; thus, we can build on it the next day.

At the limit of this system we will have warm lake, cold lake, and sea waters that do not mix anywhere. What is left of the land will be a *curve* such that the warm lake, the cold lake, and the sea waters will be however close we like from any point of c. In other words, all three regions — the warm lake with its canal, the cold lake with its canal, and the sea with its canal — will border c along the entire curve. ■

Certainly the curve c described in this example is very complicated. But this example may sow a seed of doubt in your mind: perhaps a non-self-intersecting closed curve hides similar surprises? Of course, you may say that each graph contains only a finite number of edges, and it is possible to limit ourselves to broken lines. Indeed, you probably think that the facts are visually clear for broken lines, are they not? Let us, however, take a look at the following example. In Figure 12.3 we have a closed broken line (you can verify that by following the line with a pencil). But is it really "clear" that this closed line divides the plane into two regions? (Is it clear whether a policeman A and a bandit B are in the same region?) And why is it clear? Is it an axiom, a theorem, or a non-demonstrable fact?

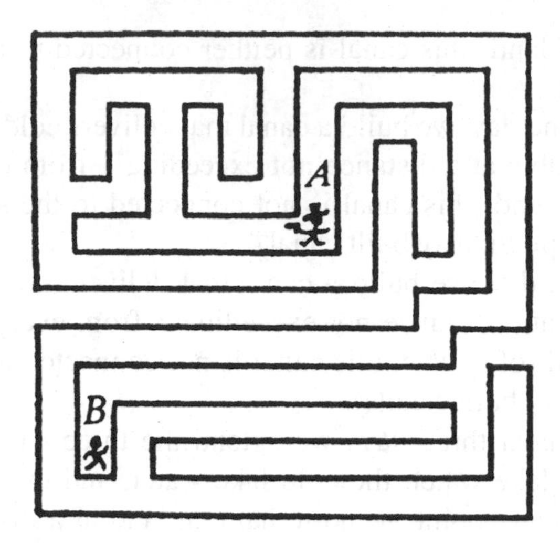

Fig. 12.3

In the nineteenth century the well-known French mathematician Camille Jordan proved the following proposition. In his theorem, Jordan considered a simple closed curve that is a continuous one-to-one image of a circle.

Jordan Curve Theorem 12.1. Any simple closed curve C in the plane divides the plane into two regions.

Every two points situated in one region can be joined by a broken line not intersecting C, whereas each broken line joining two points in different regions intersects C.

In this section we prove this theorem for an arbitrary broken contour C (like in Figure 12.3), that is, for a closed, broken, non-self-intersecting line. For this purpose we need the notion of the intersection index.

Let S and T be two segments in the plane in general position, i.e., neither of them passes through an end point of the other. If they intersect, then we write $I(S, T) = 1$; otherwise $I(S, T) = 0$. The number $I(S, T)$ is said to be the *intersection index* of segments S and T.

For convenience, any broken line in the plane (possibly with self-intersections) will be called a *chain*. A closed chain (that is, a closed broken line) will be called a *cycle*. Let X be a chain formed by segments $S_1, \ldots, S_m$, and Y be a chain formed by segments $T_1, \ldots, T_m$. Suppose that X and Y are in general position, that is, *every* segment

S_i is in general position with respect to every segment T_j. If the sum $\sum_{i,j} I(S_i, T_j)$ $(i = 1, \ldots, m; j = 1, \ldots, n)$ is *even,* then we write $I(X, Y) = 0$; otherwise, $I(X, Y) = 1$. We say that $I(X, Y)$ is the *intersection index of the chains* X and Y (more precisely, it is the intersection index modulo 2).

Exercise 12.1. Let X and Y be two cycles and L be a line not parallel to any line joining a vertex of X with a vertex of Y. We move the cycle Y continuously in the direction of L. Prove that the intersection index of the cycle X with the moving cycle Y does not change, that is $I(X, Y_1) = I(X, Y_2)$ for any two positions Y_1 and Y_2 of Y that are in general position with X.

Exercise 12.2. Prove that the intersection index of any two cycles X, Y in general position is equal to 0.

Example 12.2. Using the intersection index, we can prove (in a manner distinct from the proof in Section 11) that the graph K_5 is not planar.

Indeed, we draw all the edges of the graph K_5 (with intersections) as broken lines in general position with respect to each other, and denote by I the sum of the numbers of intersection points over all pairs of *nonadjacent* edges, taken modulo 2. Suppose we change the position of one edge, say a_1a_2 (Figure 12.4). Let X_0 be the original position of a_1a_2, and X_1 be its new position. Then the union of these chains is a cycle; we denote it $X_0 + X_1$. Further, the union

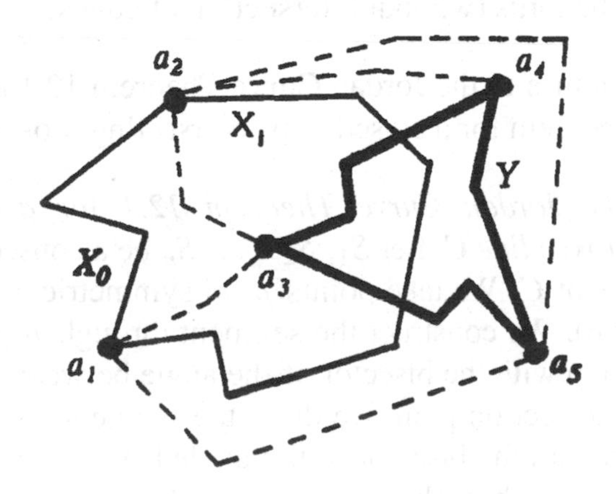

Fig. 12.4

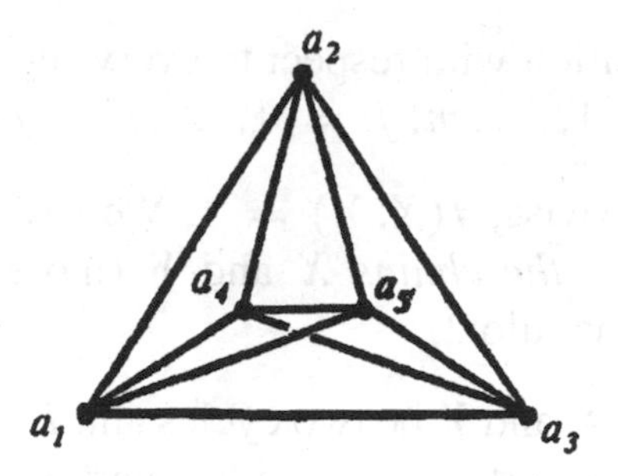

Fig. 12.5

of the edges a_3a_4, a_4a_5, a_5a_3 (that is, the edges nonadjacent with a_1a_2) is a cycle, too. We denote it Y. We know (Exercise 12.2) that $I(X_0 + X_1, Y) = 0$; therefore, $I(X_0, Y) = I(X_1, Y)$. This means that changing the position of *one* edge does not change the value of I.

But changing the positions of edges one by one we can pass from one drawing of K_5 to any other. So, the value of I is *the same* for all drawings of the graph K_5. But for the drawing in Figure 12.5 we have $I = 1$. Consequently, $I = 1$ for *every* drawing of the graph. This means that it is impossible to draw K_5 without intersecting edges, that is, this graph is non-planar. ■

Exercise 12.3. Using the intersection index, prove that the graph $K_{3,3}$ is non-planar.

Exercise 12.4. Prove that there exist two cycles X, Y on the torus such that $I(X, Y) = 1$. Does this mean that K_5 (or $K_{3,3}$) can be embedded in the torus (without intersection of edges)?

Now let us return to the Jordan Curve Theorem 12.1 and present a proof of this theorem for a closed non-intersecting broken line C.

Proof of the Jordan Curve Theorem 12.1 for a closed non-intersecting broken line **C**. Let $S_1, S_2, \ldots, S_n$ be a consecutive listing of all segments of C. We take points p, p' symmetric with respect to S_1 (Figure 12.6). We construct the segment through p parallel to S_1 to its intersection with the bisector of the angle between S_1 and S_2.

From the intersection point we draw the segment parallel to S_2 to its intersection with the bisector of the angle between S_2 and S_3, and so on. We end up with the broken line M, whose segments are all the

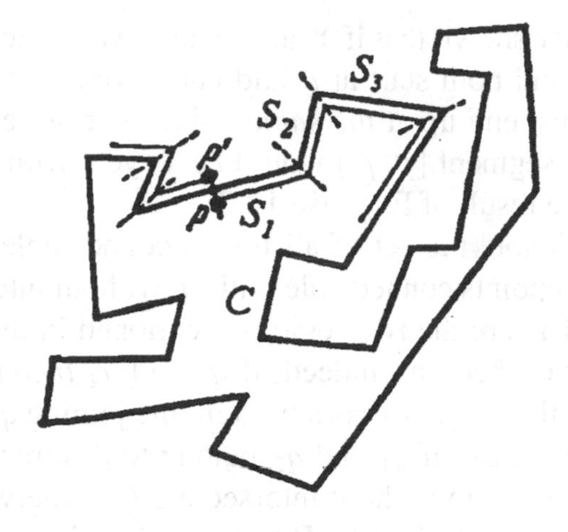

Fig. 12.6

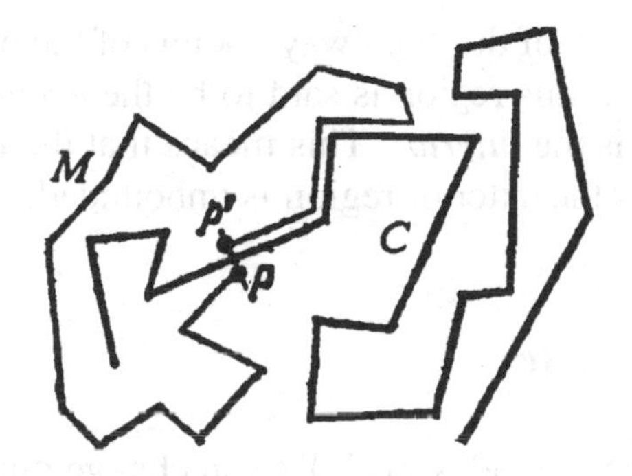

Fig. 12.7

same distance from the corresponding segments of the initial broken line C. If $|pp'|$ is small enough, C and M won't intersect. Consequently, while circling C, we see that M must return either to p or to p'. *But M cannot come to p', because that would mean the union of M and the segment $[p, p']$ is a cycle Y* (Figure 12.7) with $I(C, Y) = 1$, contradicting the result of Exercise 12.2. Thus, M is a closed broken line that goes once around C and returns to p.

Similarly, we construct the broken line M' by starting from p', going around C and returning to p'.

Now let q be an arbitrary point that does not belong to C. Then, without intersecting C, it is possible to join q with either p or p'. Indeed, let us draw a ray emanating from q that intersects M and M'. Now we can go along this ray from q to *the first* point of intersection with either M or M' and then along this broken line until we reach either p or p'.

It can be easily shown that if Y and Z are two broken lines that do no intersect C and both start at q and end at one of the points p or p', then they must end up at *the same* point (otherwise, the union W of Y, Z and the segment $[p, p']$ would be a cycle with $I(W, L) = 1$, contradicting the result of Exercise 12.2).

Let us now denote the set of all points connectable with p by U and the set of all points connectable with p' without intersecting C by V. Then U and V are the two regions mentioned in the statement of the Jordan Curve Theorem. Indeed, if q_1 and q_2 *belong to the same region (say, U),* then there exists a broken line joining q_1 and q_2 without intersecting C. But if q_1 and q_2 belong to different regions, it is impossible to join them without intersecting C (otherwise we would obtain, as above, a cycle X with $I(X, C) = 1$). We are done. ■

Let us note that all of the "far away" points of the plane are situated in the same region; this region is said to be the *exterior* region of C. The other region is the *interior*. This means that the interior region is bounded, whereas the exterior region is unbounded.

Solutions to Exercises

12.1. The intersection index $I(X, Y)$ can change only when a vertex of one cycle coincides with an interior point of a segment of the other cycle. (We note that a vertex of Xcannot coincide with a vertex of the moving cycle Y by virtue of the choice of L.) However, in those moments the parity of the number of intersection points does not change (Figure 12.8). Thus, the intersection index does not change. ■

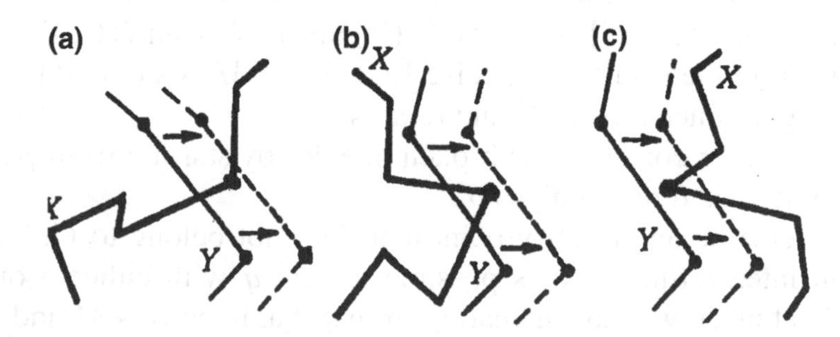

Fig. 12.8

12.2. Let us move the cycle Y, as in Exercise 12.1, from an initial position Y_0 to a position Y_1 so that Y does not intersect X (Figure 12.9). Then at the intersection index is equal to 0 position Y_1. Hence, due to Exercise 12.1, for $Y = Y_0$ we have $I(X, Y_0) = I(X, Y_1) = 0$. ■

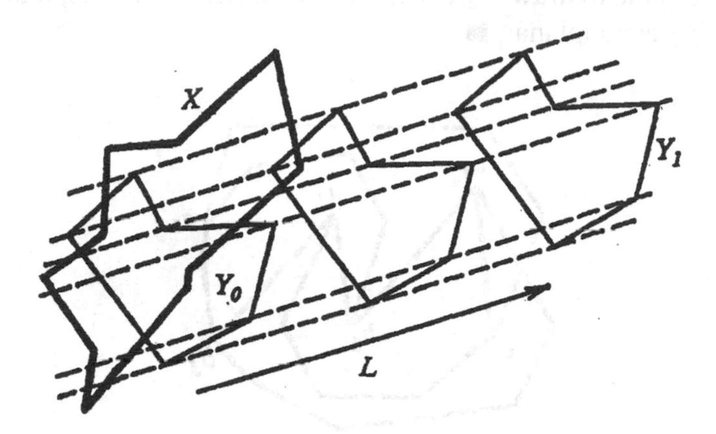

Fig. 12.9

12.3. We draw all edges as broken lines in general position (possibly with intersections) and denote the sum modulo 2 of the numbers of intersection points on all pairs of *non-adjacent* edges by I. Suppose that we change the position of one edge, say a_1b_1. We denote the initial position of the edge by X_0 and the new position by X_1 (Figure 12.10). The union of those edges is a cycle $X_0 + X_1$. Four edges a_2b_2, a_2b_3, a_3b_2, a_3b_3, are nonadjacent to a_1b_1. They form a cycle, too; we denote it Y. By Exercise 12.2, we have $I(X_0 + Y_1, Y) = 0$; therefore $I(X_0, Y) = I(X_1, Y)$. This means that changing the position of *one*

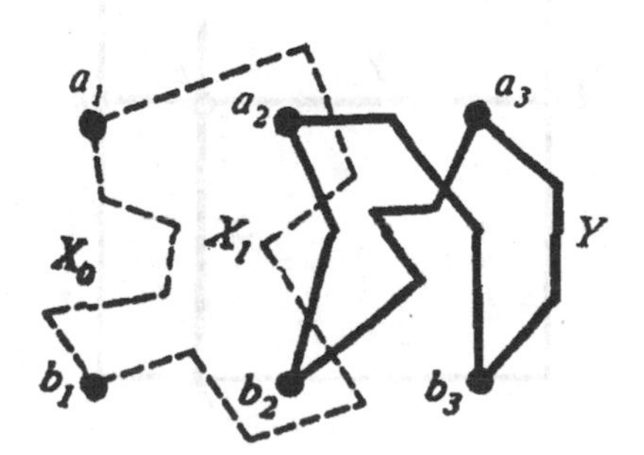

Fig. 12.10

edge does not change the value of I. By changing one by one the positions of the edges, we conclude that the value of I is *the same* for all drawings of $X_{3,3}$. But for the drawing in Figure 12.11 we have $I = 1$. Consequently, $I = 1$ for *every* drawing of the graph. This means that it is impossible to draw $X_{3,3}$ without intersection of edges, that is, the graph $X_{3,3}$ is nonplanar. ■

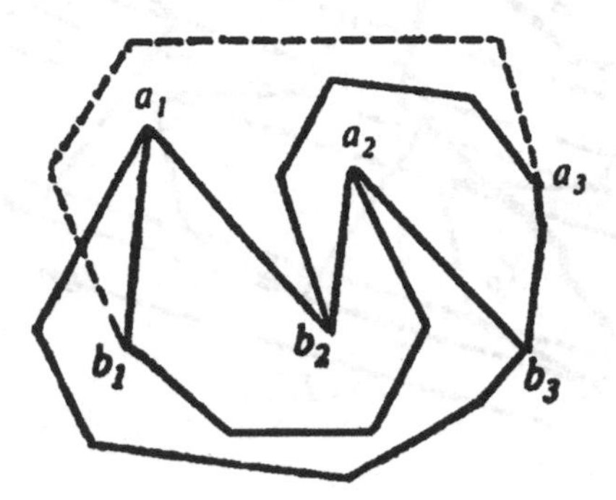

Fig. 12.11

12.4. In Figure 12.12, two cycles X and Y are drawn on the torus that have only one intersection point C (see p. 50 for the definition of a torus). Thus, $I(X, Y) = 1$. This means that the reasoning of Example 12.2 and the solution to Exercise 12.3 fail in the case of a torus. This gives us hope that the graphs X_5 and $X_{3,3}$ can be embedded in the torus (without intersection of edges). Indeed, an embedding of

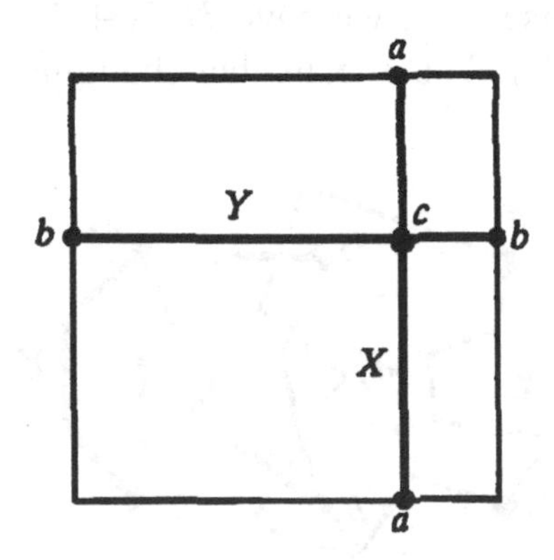

Fig. 12.12

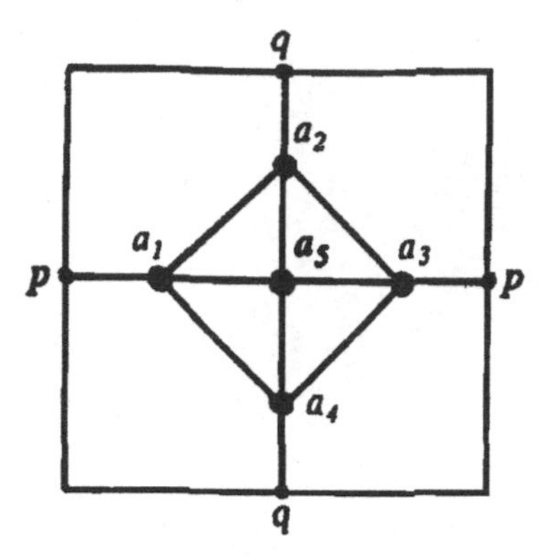

Fig. 12.13

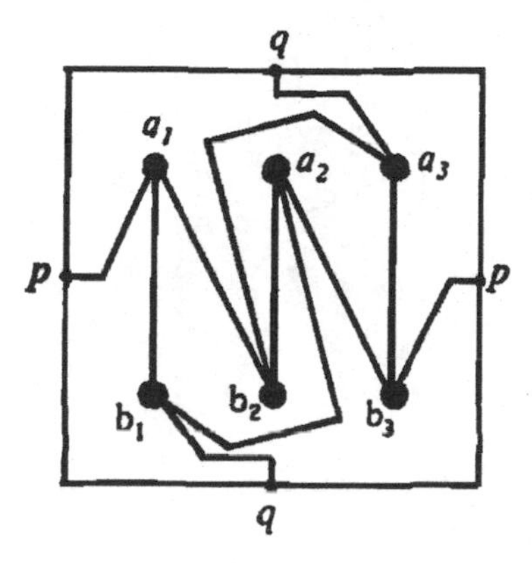

Fig. 12.14

K_5 is shown in Figure 12.13, and an embedding of $K_{3,3}$ is shown in Figure 12.14 (see also Figure 12.11). But it should be noted that there exist graphs that cannot be embedded in the torus (without intersection of edges). ■

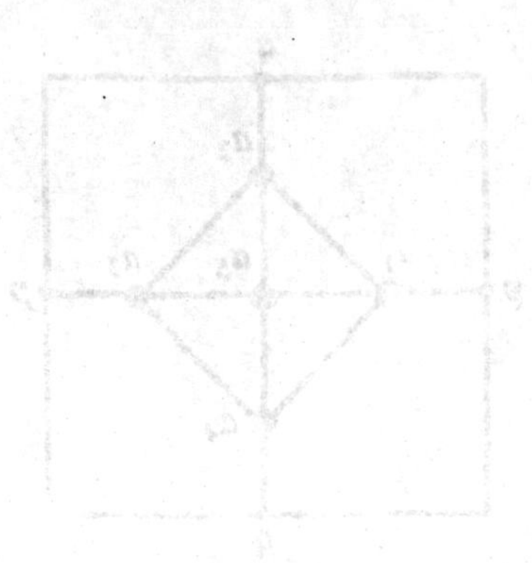

Chapter 4
Ideas of Combinatorial Geometry

13 What are Convex Figures?

In this section we discuss (without proof) some properties of convex figures in the plane.

A figure M is called *convex* if, together with every two points a, b, it contains the whole segment $[a, b]$, (Figure 13.1). Triangles, parallelograms, and trapezoids are convex (Figure 13.2). Disks, sectors with a central angle not exceeding 180°, and segments of disks are also convex (Figure 13.3); so is any segment and any single point. Some nonconvex figures are shown in Figure 13.4.

All the figures mentioned above are bounded: each is contained in a large enough disk (or square).

A *strip* is an example of an unbounded convex figure (Figure 13.5). Angles may be convex or nonconvex. If an angle's measure does not exceed 180°, then the angle is convex (Figure 13.6); otherwise, it is nonconvex (Figure 13.7).

A point a of a triangle T is called a *boundary point* if any disk with center a contains a point x that belongs to T as well as a point y that does not belong to T (Figure 13.8). A point a of a triangle T is its boundary point if and only if a belongs to one of the sides of T. All other points of T are *interior points*. For each interior point b of T (Figure 13.9), there exists a disk D with center b such that D is completely contained in T (it maybe very small). The above definitions are related not only to a triangle, but to any convex figure: any of the convex figure's points can be either a boundary point (Figure 13.10) or an interior point (Figure 13.11).

A figure is *closed* if it contains all its boundary points. *In this and all following sections a "figure" will mean "a closed figure."* A convex

A. Soifer, *Geometric Etudes in Combinatorial Mathematics*,
DOI 10.1007/978-0-387-75470-3_4,

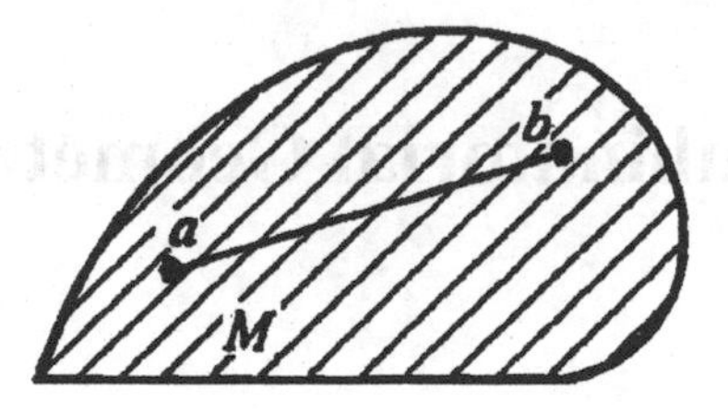

Fig. 13.1

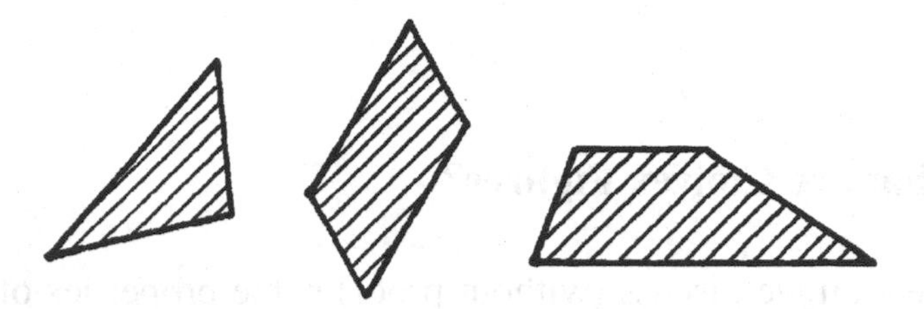

Fig. 13.2

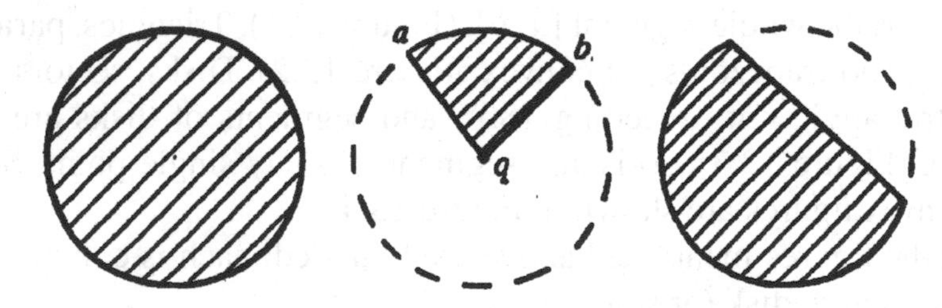

Fig. 13.3

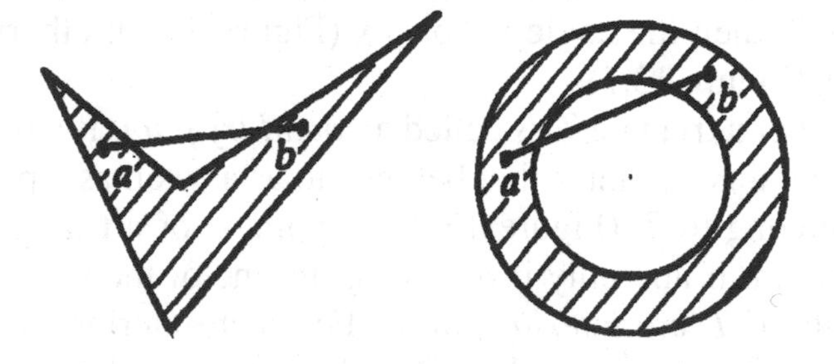

Fig. 13.4

figure that is closed and bounded is called *compact*. We will denote the *interior* of a figure M by *intM*, that is, the set of all interior points of M, and its *boundary* by *bdM*, that is, the set of all boundary points of M.

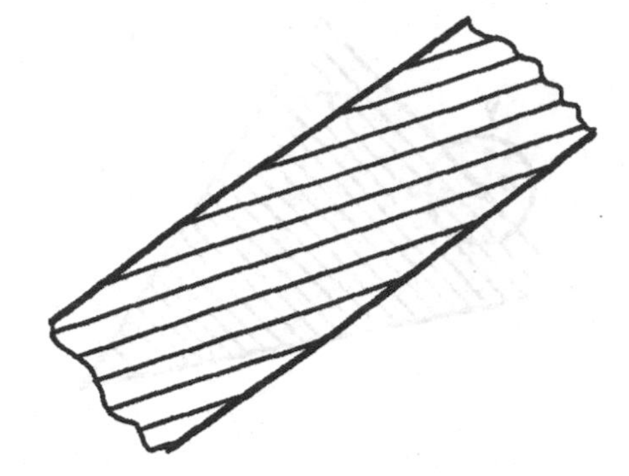

Fig. 13.5

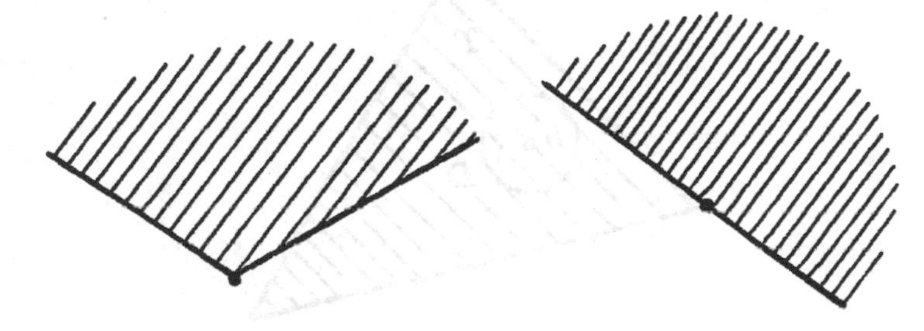

Fig. 13.6

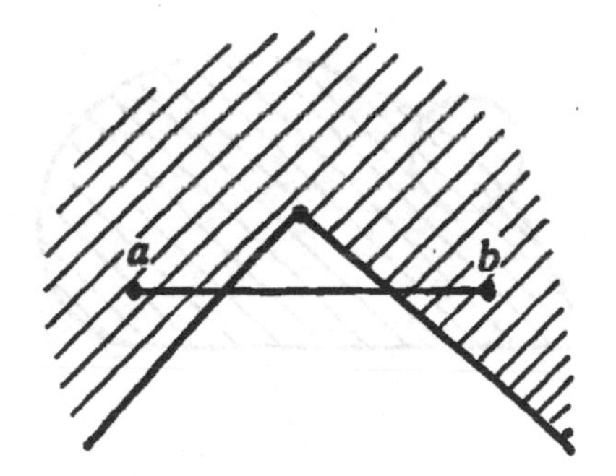

Fig. 13.7

Exercise 13.1. Let a be an interior point of a convex figure M and p a boundary point of M. Prove that all the points of the segment $[a, p]$ except p are interior points of M.

Exercise 13.2. Prove that the intersection of two or more convex figures is a convex (possibly empty) figure.

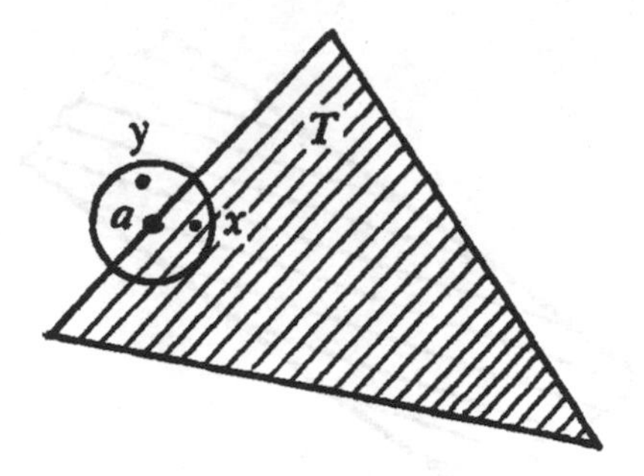

Fig. 13.8

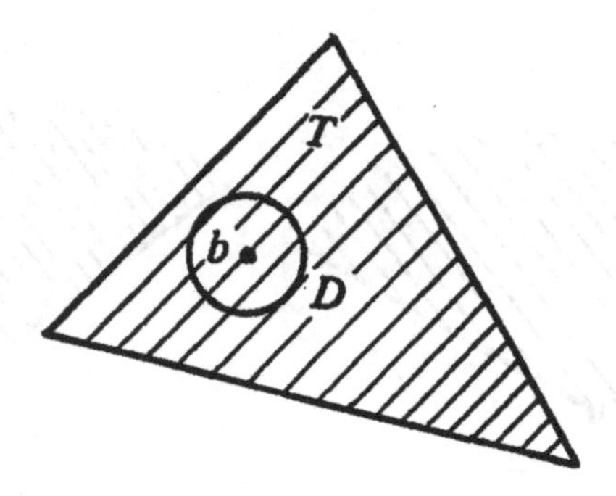

Fig. 13.9

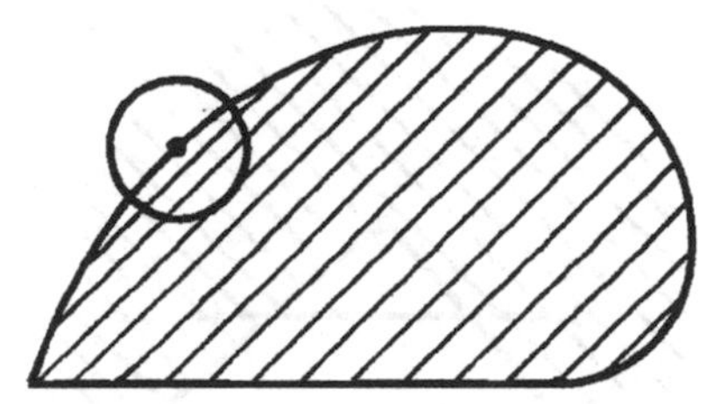

Fig. 13.10

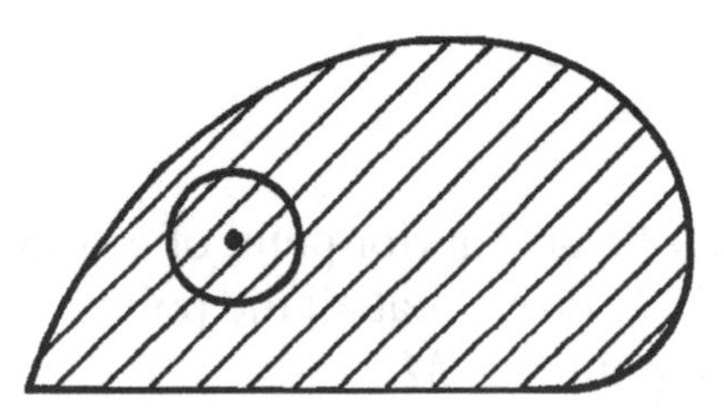

Fig. 13.11

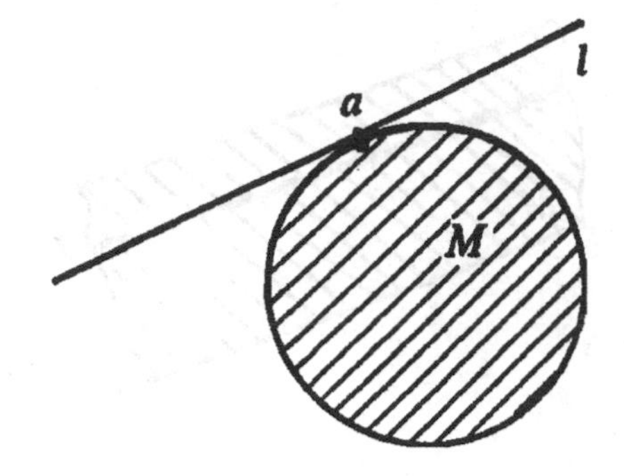

Fig. 13.12

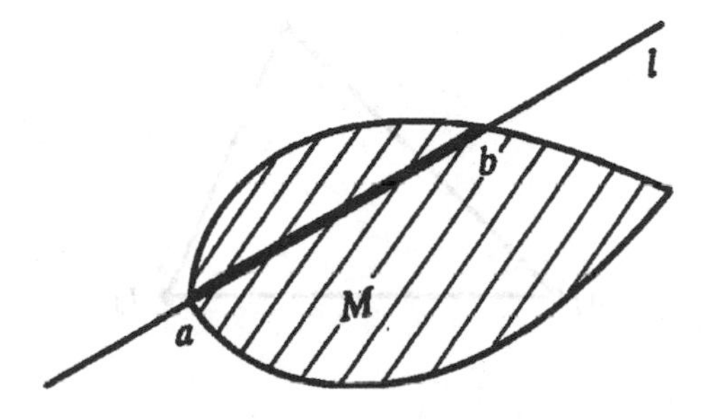

Fig. 13.13

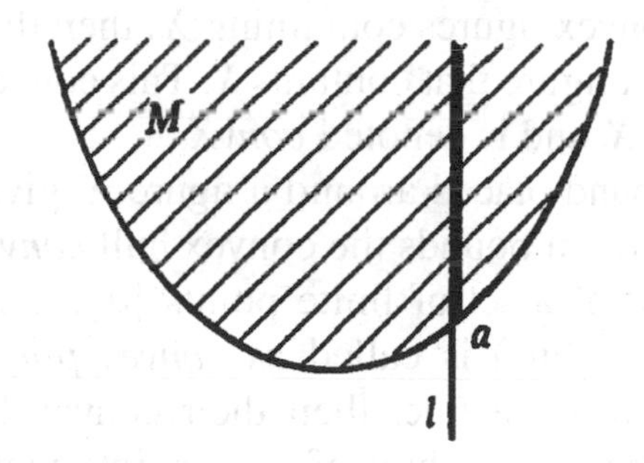

Fig. 13.14

Every line is a convex figure. Consequently, the intersection of line l with a convex figure M is a convex figure situated on the line l, that is either a point, segment, ray, or the entire line (Figures 13.12–13.15).

Exercise 13.2 implies that for every figure X (not necessarily convex) there exists the *smallest* convex figure that contains X. Indeed, if

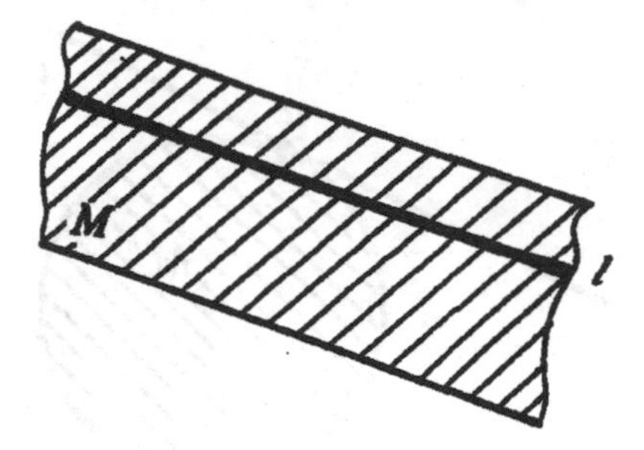

Fig. 13.15

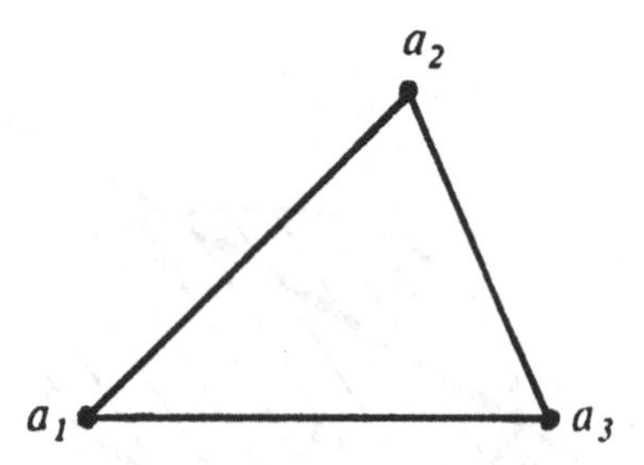

Fig. 13.16

we consider *all* convex figures containing X, then their intersection is the smallest convex figure that contains X. This convex figure is called the *convex hull* of X and is denoted *conv*X.

A tight rubber band placed around a figure X gives a good visualization of this notion: it bounds the convex hull *conv*X *of* X.

The convex hull of a set of finite points $\{a_1, \ldots, a_s\}$ in the plane (not all points on a line) is called as *convex polygon*. If we have three points, not all on a line, then their convex hull is a triangle (Figure 13.16). The convex hull of four points may be either a triangle (Figures 13.17 and 13.18) or a quadrangle (Figure 13.19). In general, if we consider the convex hull of a finite set $\{a_1, \ldots, a_s\}$, then it is possible to permute subscripts and choose a minimal subset, say $\{a_1, \ldots, a_k\}$ $(k \leq s)$, such that $conv\ \{a_1, \ldots, a_s\} = conv\{a_1, \ldots, a_k\}$. This means that the points $a_{k+1}, \ldots, a_s$ are situated on the boundary or in the interior of the convex polygon $M = conv\{a_1, \ldots, a_k\}$ (Figure 13.20).

Similarly, the convex hull of a finite point set in space (not all points in a plane) is called a *convex polyhedron*.

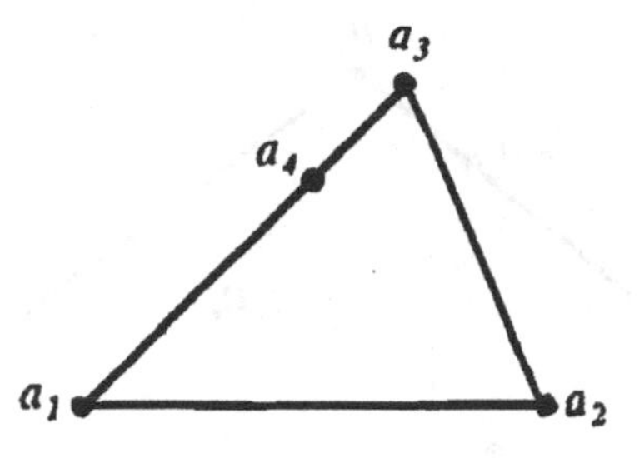

Fig. 13.17

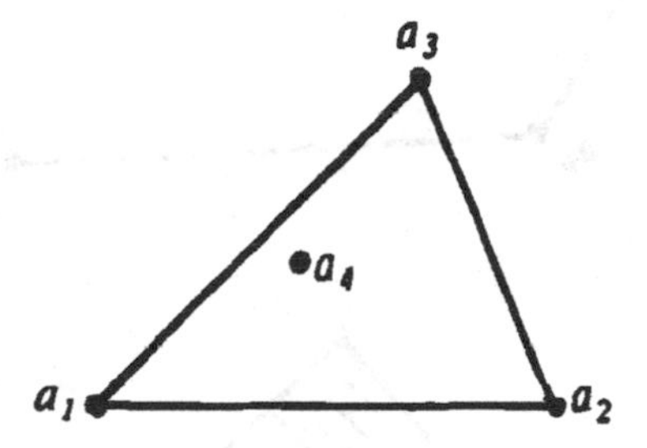

Fig. 13.18

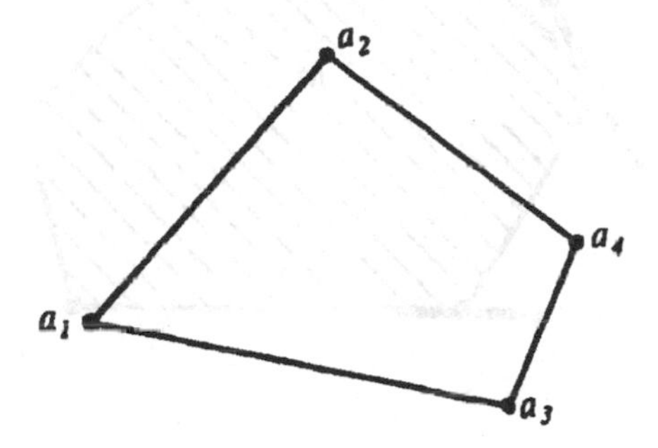

Fig. 13.19

If $M = conv\{a_1 \ldots a_m\}$ is a convex polygon and the set $\{a_1, \ldots, a_m\}$ is minimal, then none of the points $a_1, \ldots, a_m$ belongs to the convex hull of all other points. These points are called *vertices* of the convex polygon M. Each vertex a_j of the convex polygon M is an *extremal point*, i.e., there exists no segment $[p, q]$ contained in M, such that a_j is an interior point of this segment (Figure 13.21). All other points of M are non-extremal (Figure 13.22). So, each convex polygon is the convex hull of the set of its extremal points (and this set is minimal). This property of convex polygons has a generalization for

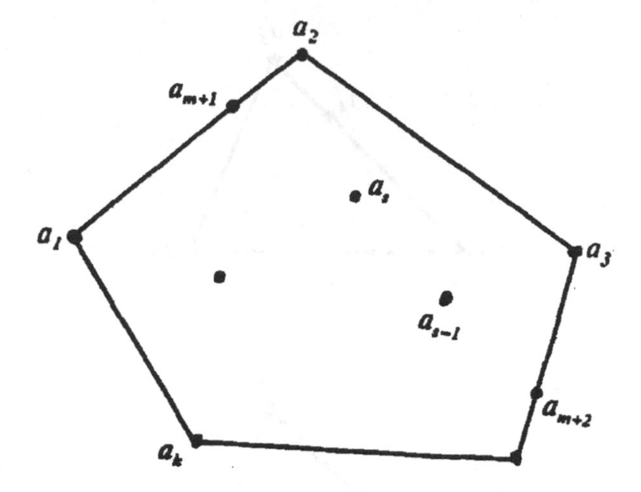

Fig. 13.20

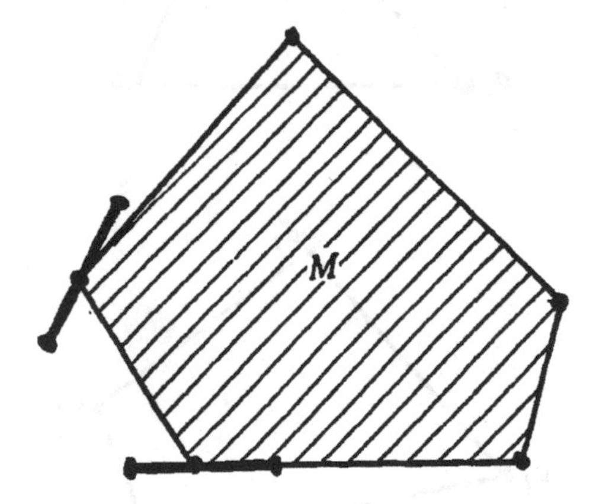

Fig. 13.21

all compact convex figures: each compact convex figure is the convex hull of the set of its extremal points. For a sector (Figure 13.3), the set of extremal points consists of its center q and all points of the arc ab.

Let F be a convex figure and L a line in the plane. We call L a *support line* of F if:

i) L and F have at least one common point and
ii) F is situated in a closed half-plane defined by L (Figure 13.23).

If F is compact then the intersection of F and its support line L is either a point (Figure 13.23) or a segment (Figure 13.24).

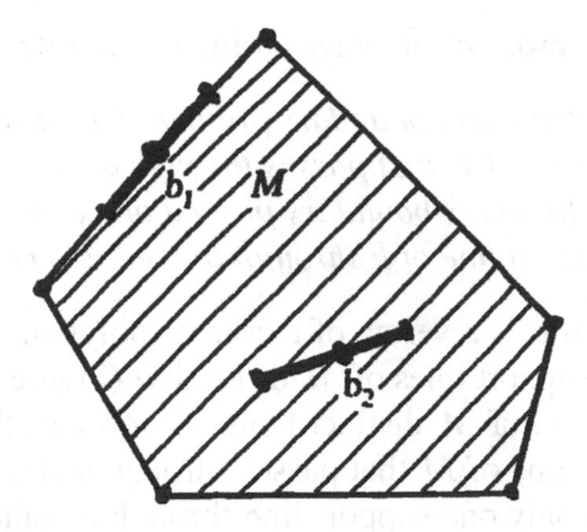

Fig. 13.22

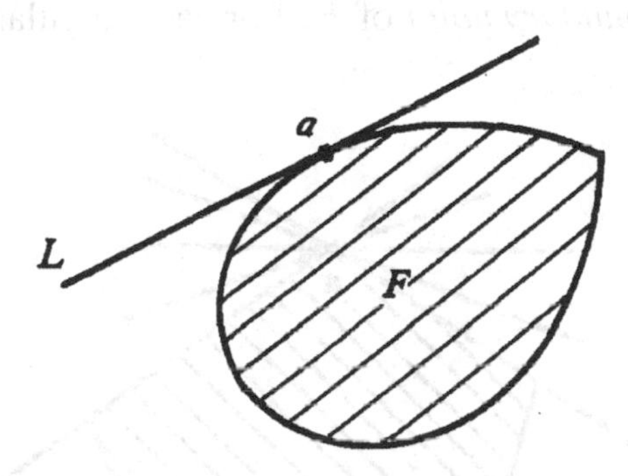

Fig. 13.23

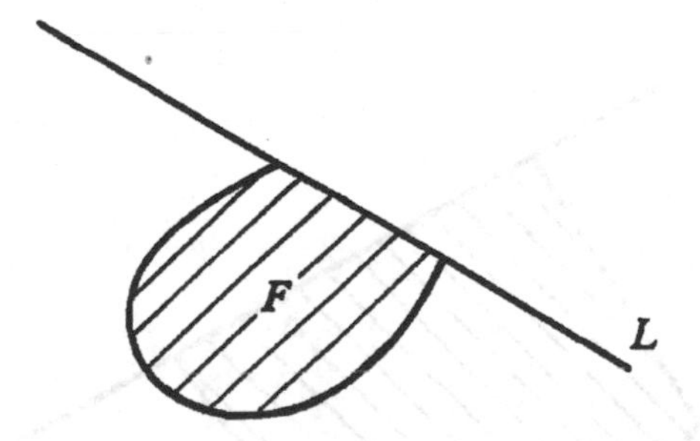

Fig. 13.24

The following proposition plays an important role in geometry.

Theorem 13.1. *For every boundary point a of a convex figure F, there exists a support line of F that passes through a.*

Conversely, if for every boundary point a of a closed plane figure F, there exists a support line of F through a, then F is convex.

For example, if a is a vertex of a convex polygon F, then there are infinitely many support lines of M through a (Figure 13.25), and if a is a boundary point of M distinct from its vertices, then there exists only one support line of M that passes through a (Figure 13.26).

If there exists only one support line through a boundary point a of a convex figure F (Figures 13.26 and 13.27), then a is said to be a *regular boundary point* of F. Otherwise, (that is, if there are infinitely many support lines through a, (Figures 13.25 and 13.28)), a is said to be an *angular boundary point* of F. For each angular boundary point

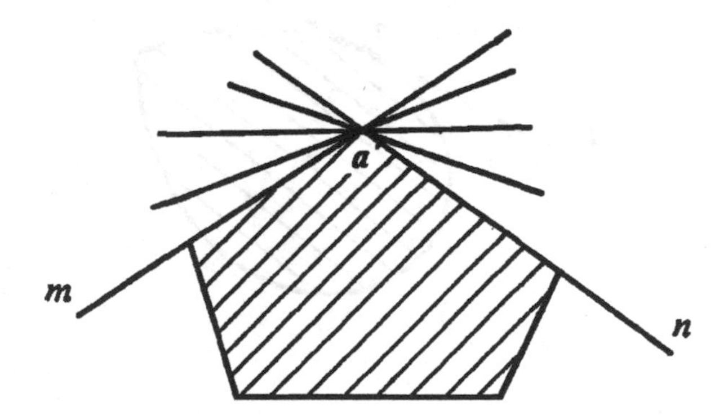

Fig. 13.25

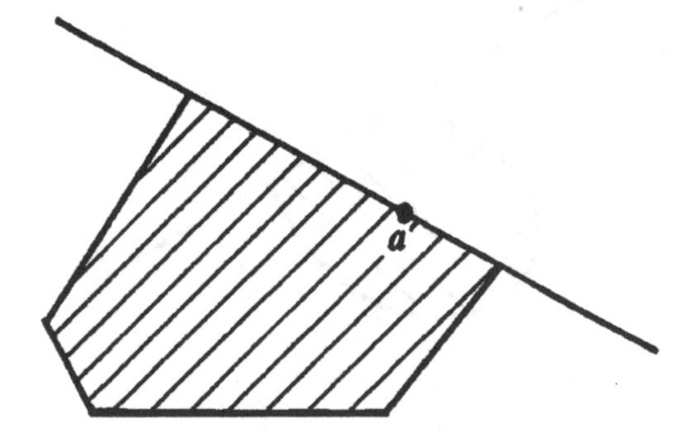

Fig. 13.26

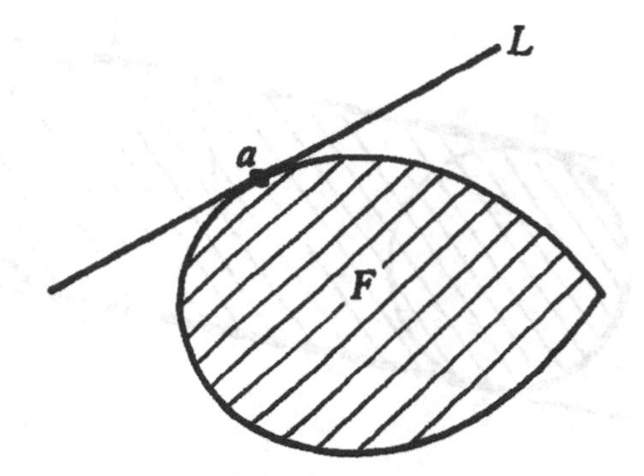

Fig. 13.27

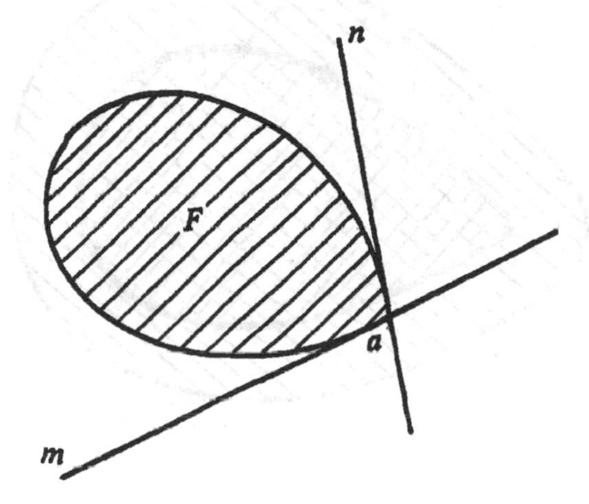

Fig. 13.28

a of a convex figure F there are two *tangent rays* of F emanating from a (the rays am and an in Figures 13.25 and 13.28). The figure F is contained in the angle between these tangent rays.

Exercise 13.3. Prove that the figure M bounded by portions of two congruent circles and their two common exterior tangent lines (Figure 13.29) is convex.

Exercise 13.4. Prove that if M is convex, then its ε-extension (Figure 13.30) is also convex (see the definition in Section 8).

Exercise 13.5. Let M be a convex polygon of perimeter d and V the ε-extension of M. What is the length of the boundary of V?

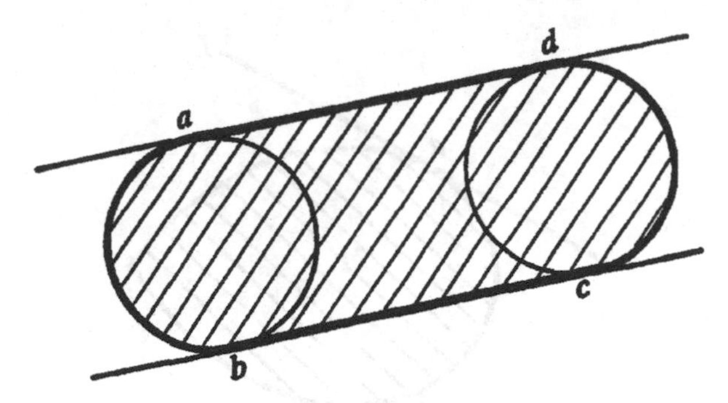

Fig. 13.29

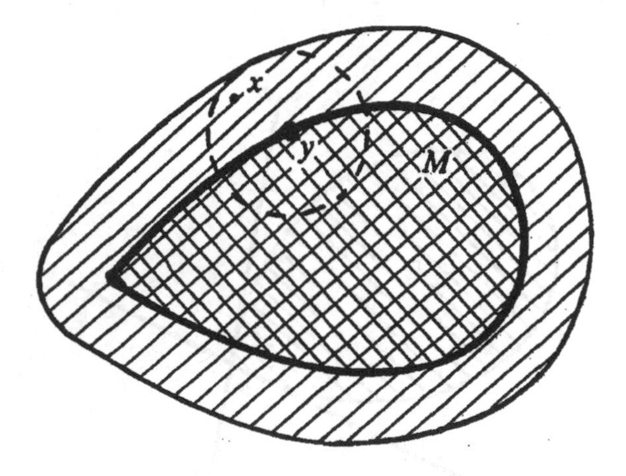

Fig. 13.30

Now we can state a very important theorem that is due to the German mathematician Wilhelm Blaschke (1885–1962):

Blascke's Theorem 13.2. Every infinite sequence of compact convex figures that is bounded contains a subsequence that converges to a compact convex figure.

Proof. The reader has probably noticed that this theorem is a corollary of Theorem 8.1. Indeed, we have to prove only that the limit M of a convergent subsequence $F_1, F_2, \ldots$ of compact convex figures is compact and convex. Let us denote the distance between the compact figures M and F_i by d_i. Then the sequence $d_1, d_2, \ldots$ of positive numbers converges to 0. For every $i = 1, 2, \ldots$ we denote the d_i-neighborhood of F_i by V_i. Then $V_1, V_2, \ldots$ are compact convex figures (see Exercise 13.4). Moreover, M is the intersection of all figures V_1, $V_2, \ldots$. Hence, M is convex and compact. ■

Exercise 13.6. Prove that for every compact convex figure F there exists a circumscribed circle, that is, a minimal circle that contains F. Is this circle unique?

Exercise 13.7. Prove that for every compact convex figure F there exists an inscribed circle, that is, a maximal circle that is contained in F. Is this circle unique?

Finally, we consider *separability* of convex figures. Convex figures F_1 and F_2 in the plane are said to be *separable* if there exists a line L such that F_1 and F_2 are situated in different closed half-planes defined by L (Figure 13.31). Let us note that F_1 and F_2 may have a common boundary point (Figure 13.31) or even a common segment (Figure 13.32). If there is a line L such that F_1 and F_2 are situated in different *open* half-planes defined by L, then F_1 and F_2 are said to be *strictly separable* (Figure 13.33).

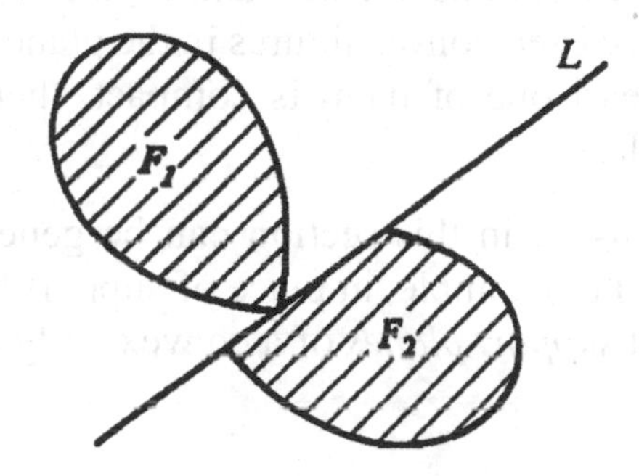

Fig. 13.31

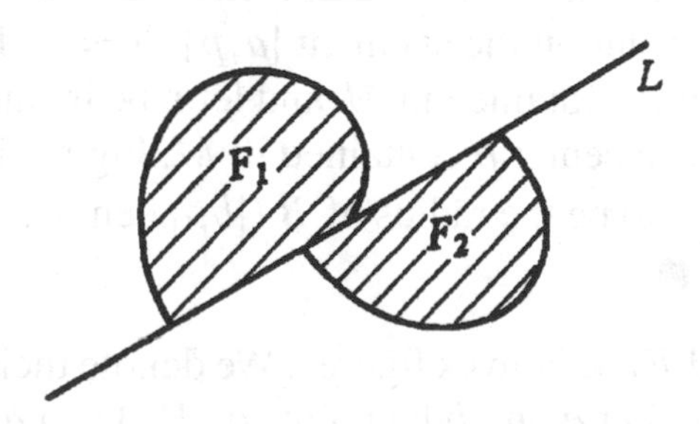

Fig. 13.32

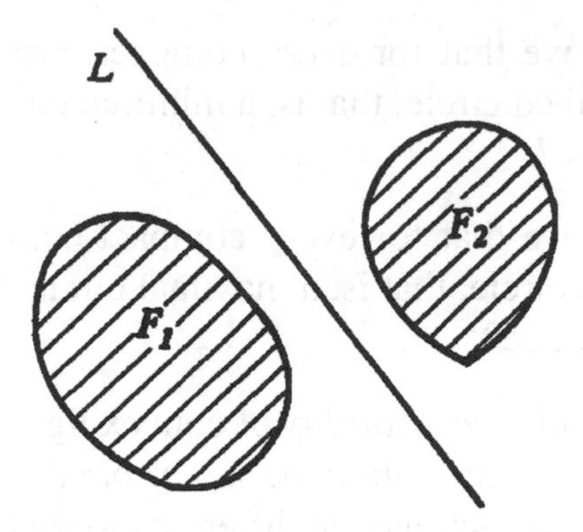

Fig. 13.33

The following propositions indicate sufficient conditions of separability for convex figures.

- If F_1 and F_2 are figures in the plane with nonempty interiors and no common interior points, then F_1 and F_2 are separable.
- If F_1 and F_2 are closed convex figures in the plane with no common points and at least one of them is compact, then F_1 and F_2 are strictly separable.

All results discussed in this section can be generalized to the n-dimensional case. For example, in place of support lines in the plane, we may talk about *support planes* of a convex body in R^3.

Solutions to Exercises

13.1. Let b be a point on the segment $[a,p]$, $b \neq p$. Further, let A be a disk with center a contained in M and let r be its radius. Then there exists a disk B with center b contained in M. Figure 13.34 shows that we may take $\frac{|bp|}{|ap|}r$ to be the radius of B ($|bp|$ denotes the length of the segment $[b,p]$). ■

13.2. Let F_1 and F_2 be convex figures. We denote their intersection by F (Figure 13.35). Let a and b be points of F. Then a and b belong to F_1 and F_2. Since F_1 is convex, all points of the segment $[a, b]$ belong to F_1. The same is true for F_2. Consequently all points of $[a, b]$ belong to F. This means that F is convex. ■

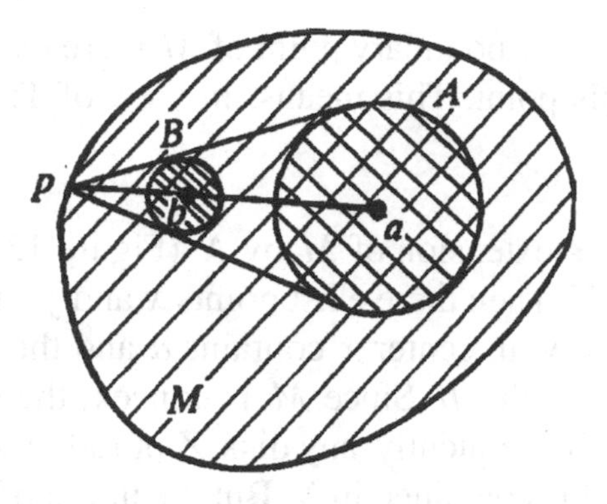

Fig. 13.34

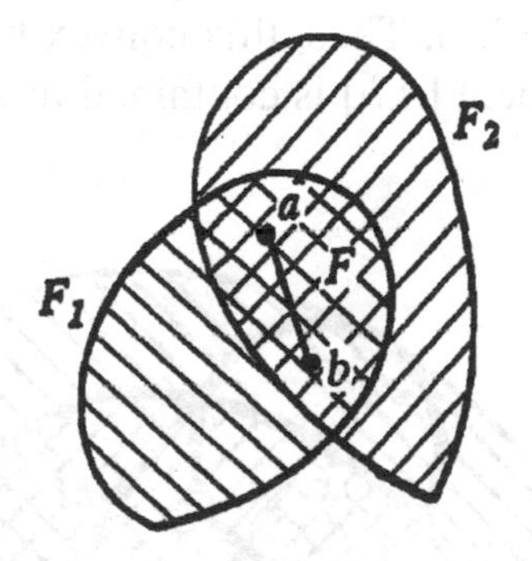

Fig. 13.35

13.3. The boundary of M consists of two semicircles ab and cd and two segments $[a,d]$ and $[b,c]$ (Figure 13.36). If x is a point of the arc ab (or of cd), then the line tangent to this arc is a support line of M. Further, if y is a point of the segment $[a,d]$, then the line containing this segment is a support line of M (similarly for segment

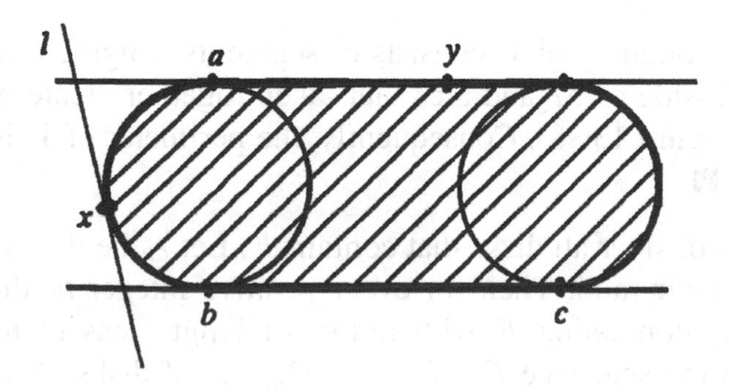

Fig. 13.36

$[b,c]$). Thus, for every boundary point of M there exists a support line of M through this point. This means, in view of Theorem 13.1, that M is convex. ■

13.4. Denote the ε-extension of M by N (Figure 13.37). Let a and b be two points of N. Then there exist points x and y in M such that the disk X of radius ε with center x contains a and the disk Y of radius ε with center y contains b. Since M is convex, the segment $[x,y]$ is contained in M. Consequently, any disk Z of radius ε with center z on the segment $[x,y]$ is contained in N. But the union of all such disks Z coincides with the convex hull of the figure consisting of two disks X and Y (see Exercise 13.3). Thus, this convex hull is contained in N. In particular, the segment $[a,b]$ is contained in N. This completes the proof. ■

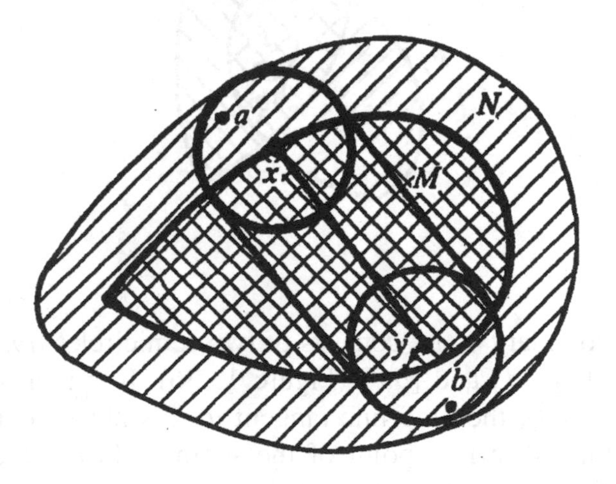

Fig. 13.37

13.5. The boundary of V consists of segments congruent to the corresponding side of M and arcs that taken together create a circle of radius ε (Figure 13.38). Consequently, the perimeter of V is equal to $p + 2\pi\varepsilon$. ■

13.6. We consider all disks that contain F. Let r_0 be the exact lower bound of their radii. Then for every positive integer m there exists a disk D_m containing F with radius of length less than $r_0 + \frac{1}{m}$. We obtain the sequence $D_1, D_2, \ldots, D_m, \ldots$ of disks. According to Blaschke's Theorem 13.2, a subsequence of this sequence converges

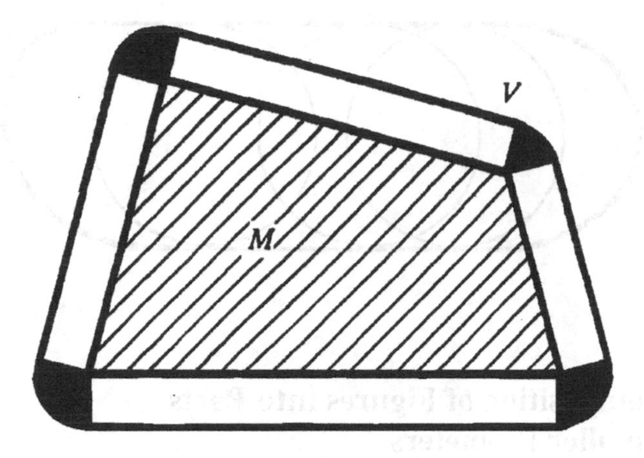

Fig. 13.38

to a compact convex figure D_0. Now it is clear that D_0 is a disk and its radius is equal to r_0. Hence, D_0 is the minimal radius disk containing F, that is, D_0 is the disk circumscribed about F.

If there were another disk D_0' of radius r_0 containing F, then F would be contained in the intersection of the disks D_0 and D_0'. But this intersection as well as the figure F is contained in a disk of radius *smaller* than r_0 (bold circle in Figure 13.39), contradicting the definition of r_0. Thus, the circumscribed disk is unique. ■

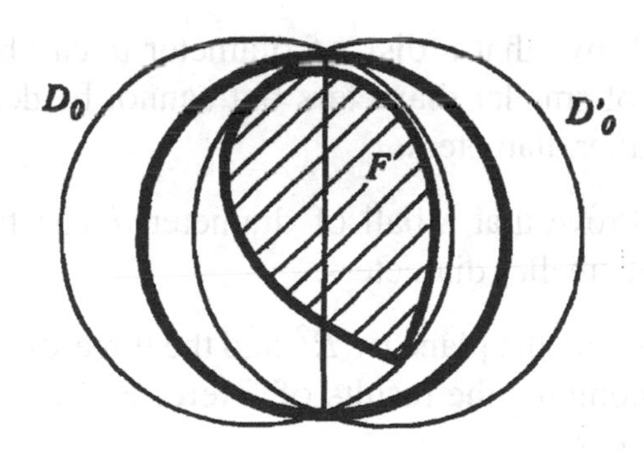

Fig. 13.39

13.7. The solution is similar to the solution to Exercise 13.6, but in this case uniqueness does not hold! This is shown in Figure 13.40. However, *if the boundary of F contains no segment, then the inscribed circle is unique.* We do not wish to deny the readers the pleasure of discovering the proof of this on their own. ■

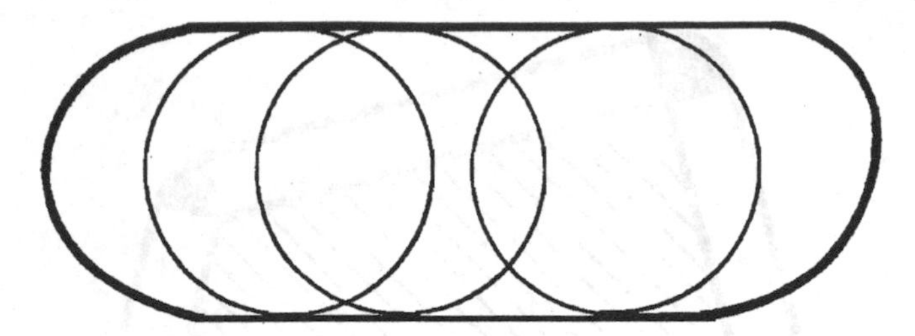

Fig. 13.40

14 Decomposition of Figures into Parts of Smaller Diameters

The *diameter* of a figure is the greatest distance between its two points. More precisely, a figure F has diameter d if

i) any two points of F are at a distance at most d and
ii) there are two points x and y in F whose distance is equal to d.

The problem of decomposition of figures into parts of smaller diameters was formulated [Bo2] by the Polish mathematician Karol Borsuk (1905–1982) in 1933. The birth of this problem was preceded by three mathematical events. The first is described in the following two exercises.

Exercise 14.1. Prove that a disk of diameter d can be decomposed into three parts of smaller diameters but cannot be decomposed into two parts of smaller diameters.

Exercise 14.2. Prove that a ball of diameter d can be decomposed into four parts of smaller diameters.

Denoting as before the plane by R^2 and the three-dimensional space by R^3, we can combine the results of Exercises 14.1 and 14.2 in the following statement:

Theorem 14.1. *The ball in n-dimensional Euclidean space R^n (where $n = 2$ or 3) can be decomposed into $n + 1$ parts of smaller diameters.*

It is not too difficult to prove (if you are familiar with the notion of n-dimensional space for arbitrary positive integer n) that this fact holds for *any* finite-dimensional Euclidean space. But Exercise 14.1 contains more information. It confirms that for $n = 2$ the

n-dimensional ball cannot be decomposed into n parts of smaller diameters. Is it true for arbitrary n? This question brought about the second mathematical event: in 1932, Karol Borsuk proved [Bo1] that for any $n \geq 2$, an n-dimensional ball of diameter d cannot be decomposed into n parts of smaller diameters. Note that the same result was obtained by the two Russian mathematicians, Lazar Lusternik and Lev Schnirelmann in 1930, before Borsuk [LS]. However, K. Borsuk did not know this and obtained the result independently (and with a different proof).

The third event occurred in 1933. Borsuk proved [Bo2] the following proposition:

Borsuk's Theorem 14.2. Every figure of diameter d in R^2 (not only the disk) can be decomposed into three parts of smaller diameter.

We are going to sketch a proof of this result, following Borsuk's ideas. He used an interesting geometrical result [P] obtained in 1921 by the Hungarian mathematician, J. F. Pál. Here is Pál's result:

Pál's Theorem 14.3. *Any plane figure F of diameter d can be inscribed in a regular hexagon with distance d between its opposite sides* (Figure 14.1).

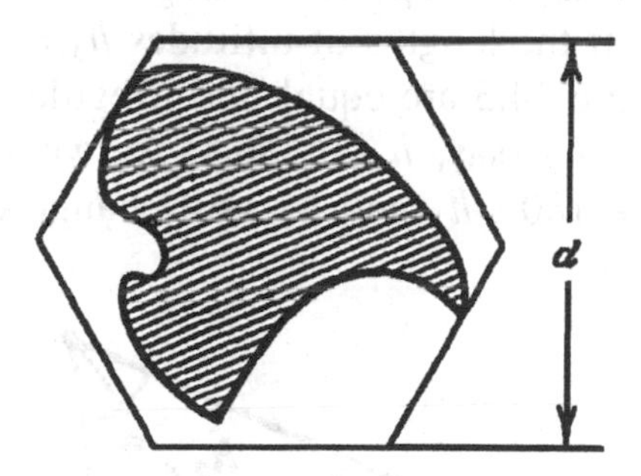

Fig. 14.1

Here is the idea of the proof: we draw an initial ray m_0 from which we may measure angles. Now we construct two parallel support lines of the figure F, L_1 and L_2, that make the angle ϕ_0 with m_0 (Figure 14.2). The distance between L_1 and L_2 does not exceed d (can you see why?). If this distance is less than d, then we move these lines away from each other until the distance between them is equal to

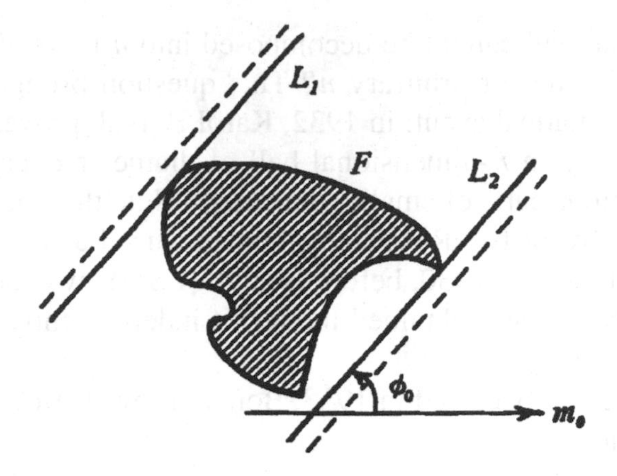

Fig. 14.2

d (dotted lines in Figure 14.2). Thus, for every angle ϕ there is a strip of width d containing F. We denote this strip by $S(\phi)$.

Now we construct the strips $S(\phi_0)$ and $S(\phi_0 + 60°)$. Their intersection is the rhombus containing F that has a 60° angle and distance d between its opposite sides (Figure 14.3). Finally, we construct the strip $S(\phi_0 + 120°)$ (dotted lines in Figure 14.3). The intersection of three strips is the hexagon $H(\phi)$ containing F with 120° angles. This hexagon is regular if the lengths of altitudes h_1 and h_2 drawn to the dotted lines in Figure 14.3 are equal and irregular otherwise. Let us assume that $h_1 \neq h_2$ (say, $h_1 < h_2$). As we increase the angle ϕ from ϕ_0 to $\phi_0 + 180°$, h_1 and h_2 will change continuously (this

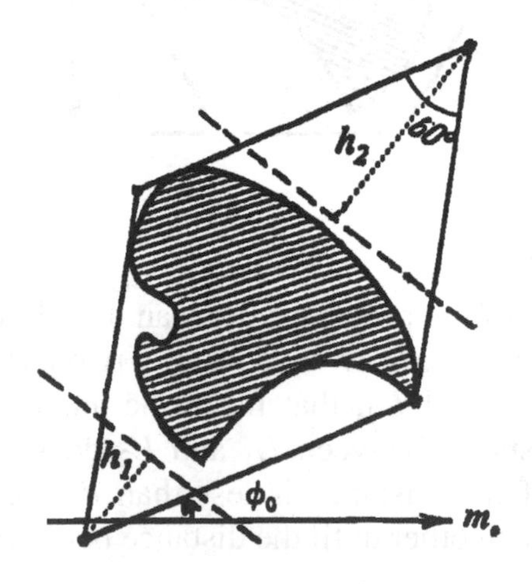

Fig. 14.3

visually clear assertion requires a rigorous proof), and consequently the difference $h_1 - h_2$ will change continuously as well. But when ϕ reaches $\phi_0 + 180°$, h_1 and h_2 will switch their roles. So, the difference, being negative at first, becomes positive in the end. Hence, by the Intermediate Value Theorem, there exists an angle $\phi = \phi_1$ such that $h_1 - h_2 = 0$. At this angle ϕ_1, the hexagon $H(\phi_1)$ is regular. ■

Exercise 14.3. Prove Borsuk's Theorem 14.5 with the help of the above result of Pál.

Let us now introduce some useful notation. For a plane figure F of diameter d we denote the smallest of the integers s such that there exists a decomposition of F into s parts of smaller diameters by $a(F)$. We will call $a(F)$ the *Borsuk number* of F.

The three events outlined above can be now combined into the following assertions:

i) $a(B) = n + 1$ for any n-dimensional ball B;
ii) $a(F) \leq 3$ for any bounded plane figure F.

These assertions led Borsuk to formulate in 1933 the following hypothesis:

The Borsuk Conjecture 14.4. For any bounded figure F in R^n,

$$a(F) \leq n + 1.$$

In fact, Borsuk was careful to ask whether the above inequality were true. But the popularity of the problem was so broad and belief in its positive answer so great that the Borsuk problem became known — in fact, famous — as the Borsuk Conjecture.

What is the state of this problem today? In 1955, the English mathematician H. G. Eggleston proved that the Borsuk conjecture is true in R^3.

Theorem 14.5. *(H.G. Eggleston [El]) Every body of diameter d in R^3 can be decomposed into four parts of smaller diameters.*

Eggleston's proof was rather complicated. In 1957, the Israeli (now American) mathematician Branko Grünbaum [G1] and the Hungarian mathematician Aladár Heppes [Hep] found alternative proofs that were much simpler. The proofs had similar main ideas: to find a "more or less small" universal cover W for all three-dimensional bodies of diameter d (this means that every body of diameter d can be embedded into W) and decompose this cover into four parts of diameters not exceeding d. So, this idea is a direct generalization of the method that was used in the proof of Theorem 14.1 (see Exercise 14.3). Grünbaum's paper was written in English and Heppes' in Hungarian; this is why the first paper is better known. Grünbaum also gives a better estimate: his four parts of the universal cover have diameters approximately $0.9887d$ whereas Heppes' estimate is approximately $0.9977d$.

Let us sketch the Grünbaum construction. As David Gale proved in [Ga], *every body of diameter d can be embedded in a regular octahedron with distance d between its opposite faces.* This fact can be proved in a manner similar to Pál's theorem. So, this octahedron is a universal cover for all bodies of diameter d. Further, Grünbaum noticed that it is possible to cut off three vertices of this octahedron by planes that are situated at distance $\frac{d}{2}$ from the center of the octahedron (see Figures 14.4, 14.5, and 14.6), and the truncated body will remain a universal cover. Finally, Grünbaum decomposes this truncated body into four parts (Figure 14.7) and computes the diameters of these parts.

And he was lucky! It turns out that if the decomposition is made carefully, then all the diameters are equal to

$$\frac{\sqrt{6129030 - 937419\sqrt{3}d}}{1518\sqrt{2}} \approx 0.9887d.$$

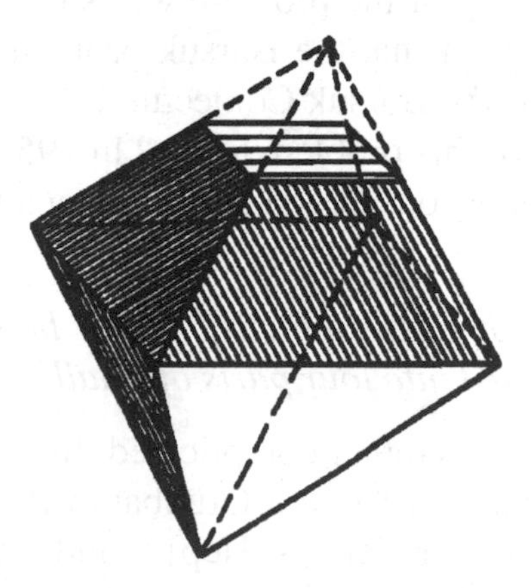

Fig. 14.4 Grünbaum's Partition, step 1

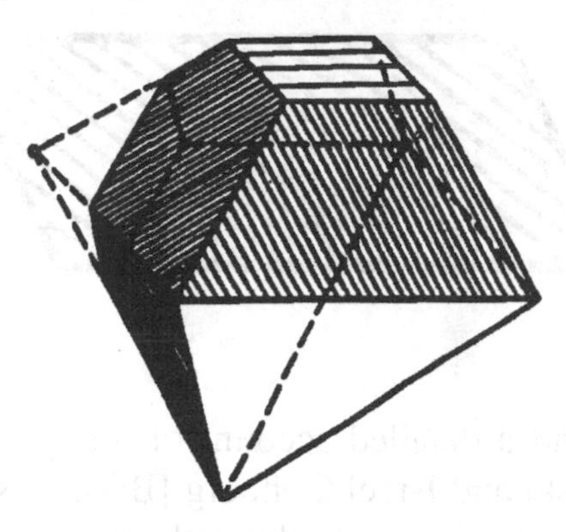

Fig. 14.5

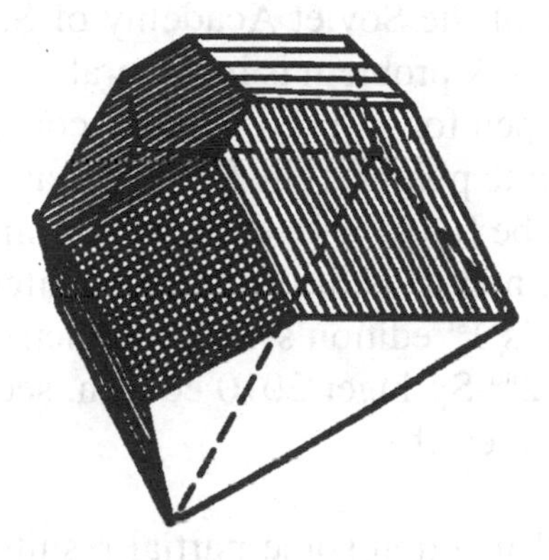

Fig. 14.6

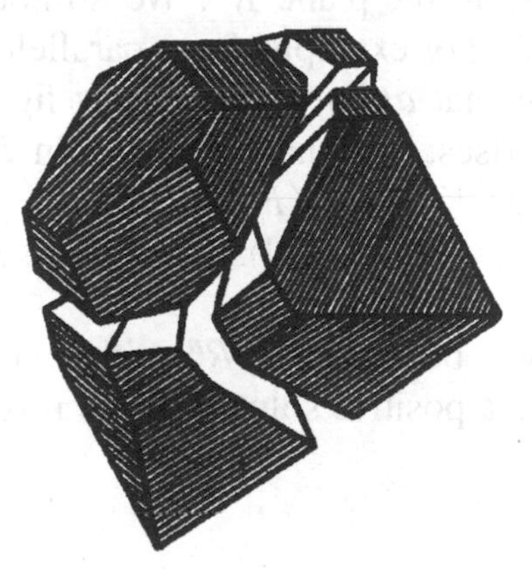

Fig. 14.7 Grünbaum's Partition, step 2

Fig. 14.8

The reader can find a detailed account of this proof in the book by Vladimir Boltyanski and Israel Gohberg [BG]. Mistaken solutions to the Borsuk problem were sent to the authors of the book [BG] many times, from different countries and cities. One such "proof" was even found by a member of the Soviet Academy of Sciences. This shows how difficult the Borsuk problem is in general.

In fact, it is still open for $n \geq 4$. Is there a convex polytope M (i.e., a convex hull of a finite point set) in R^4 such that $a(M) = 6$? Perhaps such a polytope can be found with the help of computers. At any rate, the authors are inclined to believe that a counter example exists in the case $n \geq 4$. [This 1st edition's 1990 prediction was proved to be true! In the current 2nd Springer 2010 edition, see the new Chapter 8 dedicated to this progress.]

And now we will mention some partial results in the direction of Borsuk's conjecture and its generalizations.

First, let us return to the plane R^2. We do not have $a(F) = 3$ for every plane figure F. For example, for a parallelogram F, $a(F) = 2$ (Figure 14.8). Note that $a(F) \geq 2$ for any figure F. Thus, a new problem naturally arises: for what figures F in R^2 does $a(F) = 3$? And for what figures F does $a(F) = 2$? This problem was solved by V.G. Boltyanski in 1970 ([B2]). We will discuss this solution in Section 16.

Further, for convex bodies F in R^n with a smooth boundary, the Borsuk problem has a positive solution. This result will be discussed in Section 17.

Solutions to Exercises

14.1. A decomposition of a disk into three parts of smaller diameters is shown in Figure 14.9. Another solution is given in Figure 14.10.

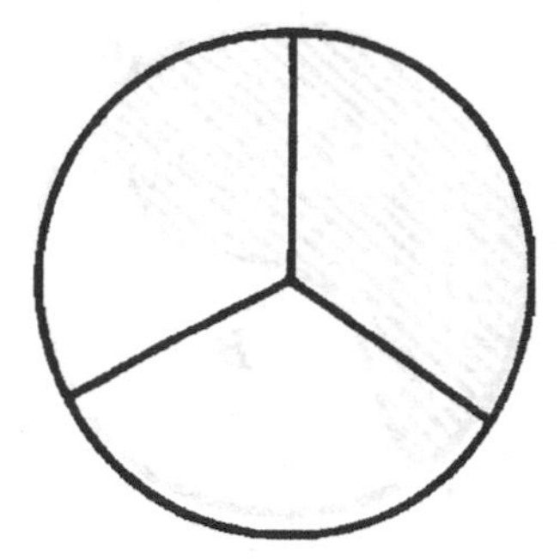

Fig. 14.9 Partitioning a disk into 3 parts of smaller diameters

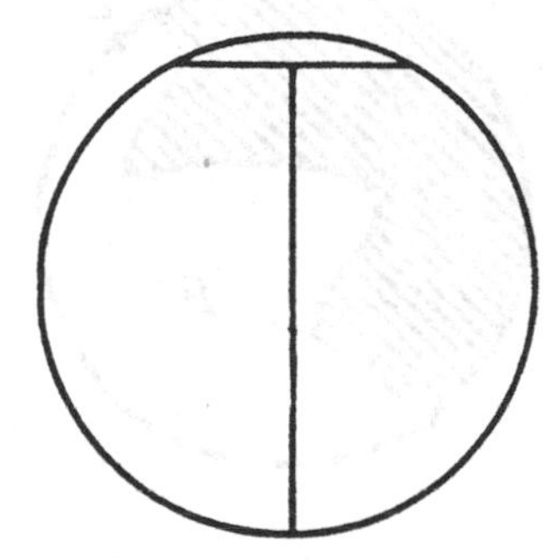

Fig. 14.10 Partitioning a disk into 3 parts of smaller diameters

Now let us assume that there exists a decomposition of the disk F of diameter d into two parts A and B of smaller diameters. We can assume without loss of generality that both sets A and B are closed because the diameter of a set is equal to the diameter of its closure. If the *boundary bdF* of F is contained in A, then the diameter of the part A is equal to d, contradicting the decomposition. So, there exists a point b in bdF that does not belong to A. That is, b belongs to B. Similarly, there exists a point a in bdF that belongs to A. Since bdF is connected and A and B are closed, there exists a point c in bdF such that c belongs to both A and B. Let now c' be a point in bdF that is diametrically opposite to c. If c' is in A, then the diameter of $A$is equal to d because c and c' both belong to A (Figure 14.11). And if c' is in B, then the diameter of B is equal to d because c and c' both belong to B (Figure 14.12). This contradiction shows that there exists no decomposition of the disk F into two parts of smaller diameters. ■

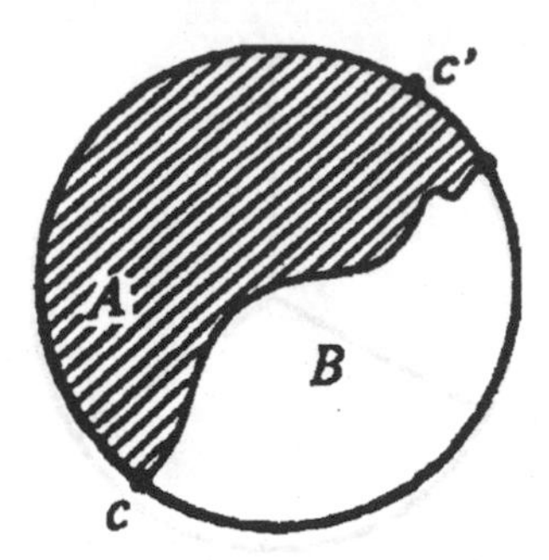

Fig. 14.11

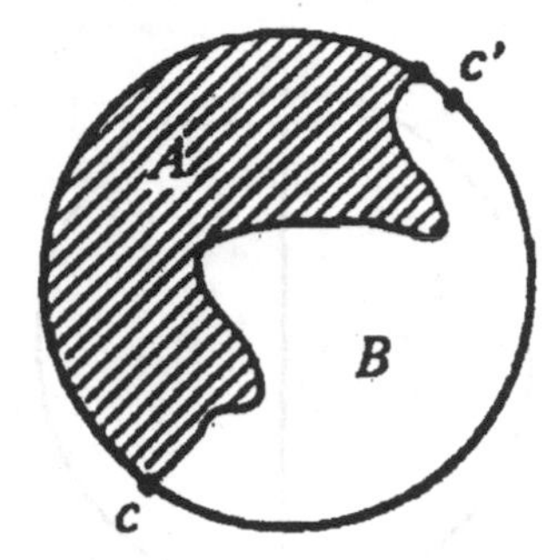

Fig. 14.12

14.2. A solution is presented in Figure 14.13(a). The upper cap is less than a hemisphere, and the three planes cutting the lower part of the ball form three dihedral angles and meet at an angle of 120°. The solution in Figure 14.13(b) is more symmetric. We inscribe a regular

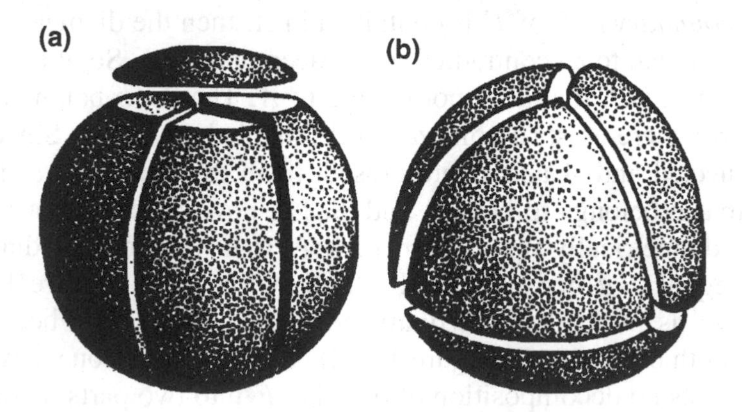

Fig. 14.13 Partitioning a ball into 4 parts of smaller diameters

tetrahedron in the ball and project its four faces onto the boundary sphere of the ball from its center. Each part of the ball is the convex hull of the center of the ball and one of the aforesaid projections. ■

14.3. A solution is shown in Figure 14.14. Indeed, the hexagon is decomposed into three parts, each of which has a diameter smaller than d. Hence, the figure F inscribed in the hexagon is decomposed, too. ■

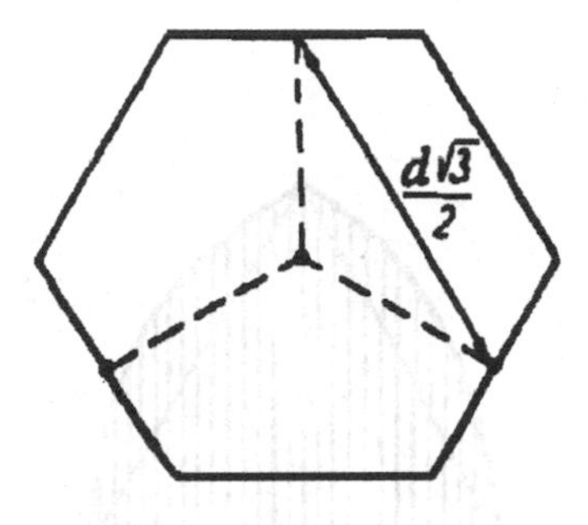

Fig. 14.14

15 Figures of Constant Width

Let F be a plane convex figure and L be a line in the plane. The distance h between two support lines of F parallel to L (Figure 15.1) is called the width of F in direction L. If the width of F is the same in all the directions, then F is said to be a *figure of constant width.*

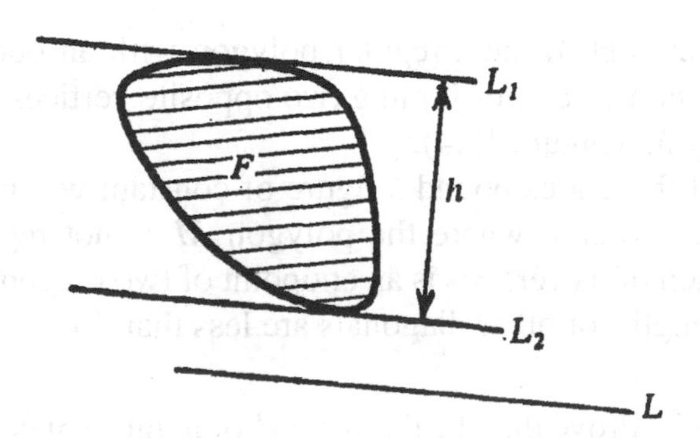

Fig. 15.1

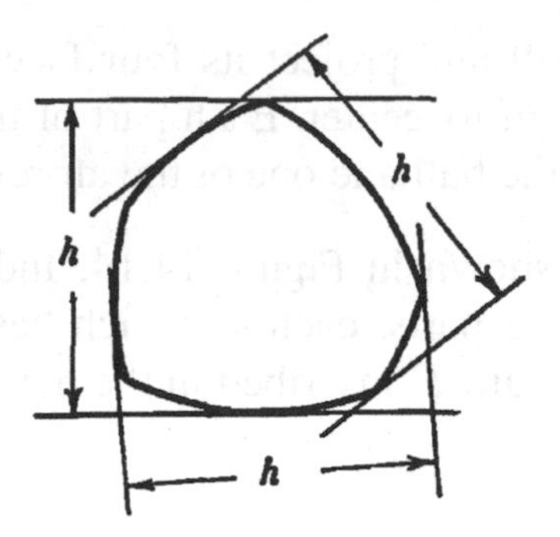

Fig. 15.2

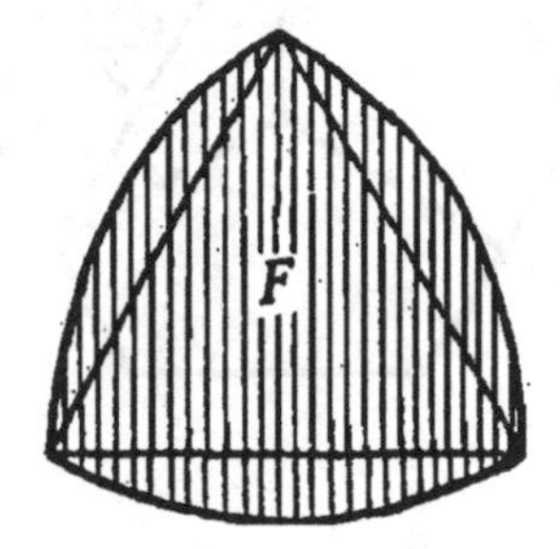

Fig. 15.3 The Reuleaux Triangle

Certainly, every disk is a figure of constant width. But there is an infinite family of figure of constant width besides disks — see Figure 15.2 for an example.

Exercise 15.1. The *Reuleaux Triangle* (Figure 15.3) is bounded by three arcs of circles with centers at the vertices of a regular triangle. Prove that this figure has a constant width.

Exercise 15.2. Let M be a regular polygon with an odd number of vertices. We construct arcs joining two opposite vertices with centers at vertices of M (Figure 15.4).

Prove that these arcs bound a figure of constant width. Generalize this result to the case where the polygon M is not regular (Figure 15.5), but each of its vertices is an endpoint of two diagonals of length h (and the lengths of other diagonals are less than h).

Exercise 15.3. Prove that the diameter d of a figure of constant width h is equal to h.

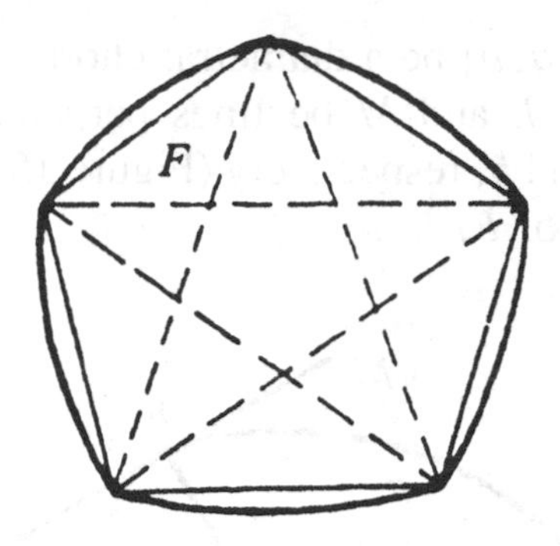

Fig. 15.4

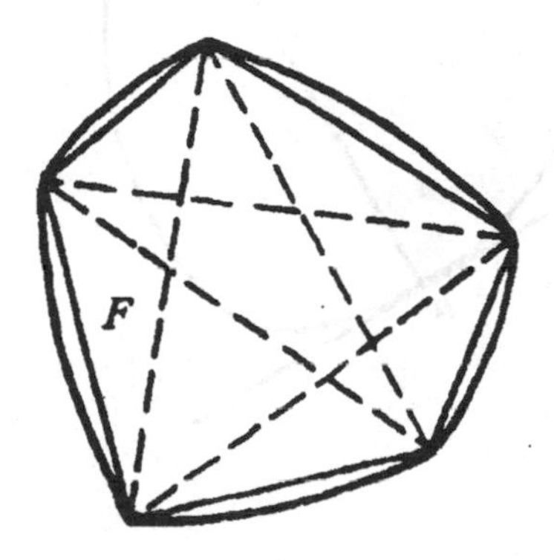

Fig. 15.5

Let F be a figure of constant width h.Every chord of F that has length h is called a *diametral chord* of F (Figure 15.6).

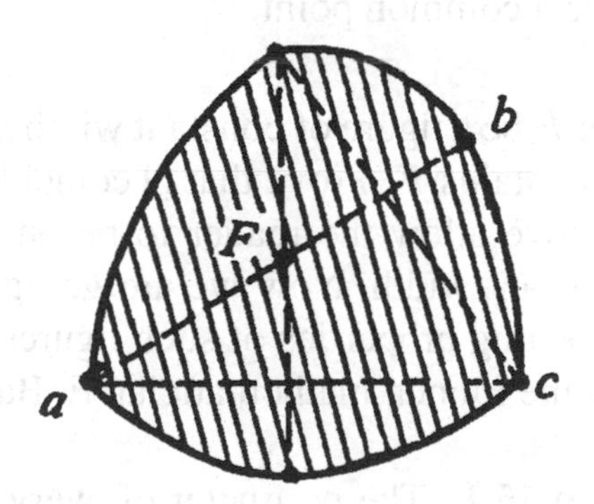

Fig. 15.6

Exercise 15.4. Prove that each boundary point a of a figure F of constant width is an endpoint of at least one diametral chord of F.

Exercise 15.5. Let $[a, b]$ be a diametral chord of a figure F of constant width, and let L and M be lines perpendicular to $[a, b]$ and passing through a and b, respectively (Figure 15.7). Prove that L and M are support lines of F.

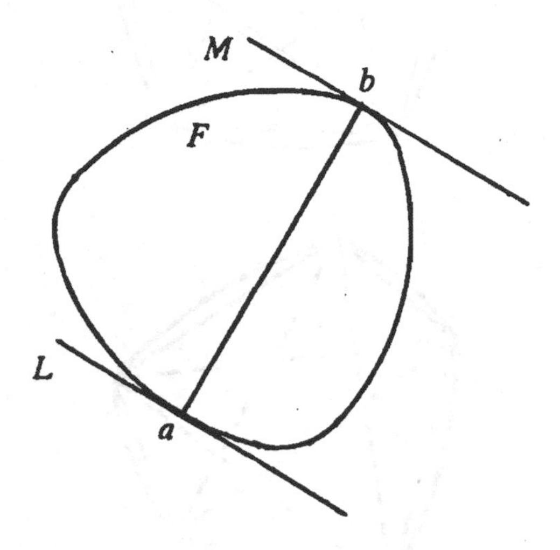

Fig. 15.7

Exercise 15.6. Prove that every support line L of a figure F of constant width has only one common point with F.

Exercise 15.7. Prove that every two diametral chords of a figure F of constant width have a common point.

Exercise 15.8. Let F be a figure of constant width h and Q be a figure of diameter h that contains F. Prove that Q coincides with F.

The exercises above allow the reader to be on a friendly footing with figures of constant width. Now we are going to consider some important and interesting properties of such figures. The first of them was established by the French mathematician E. Barbier.

Barbier's Theorem 15.1. The perimeter of every figure of constant width h is equal to πh.

Proof. First, we will formulate the following auxiliary proposition. Prove it on your own. Let $abcd$ be a rhombus and L and M be two

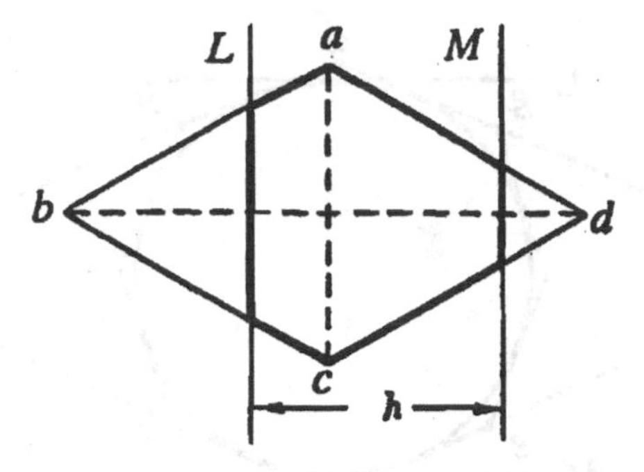

Fig. 15.8

parallel lines at distance h from each other that intersect the rhombus and are perpendicular to the diagonal $[b, d]$ (Figure 15.8). Then the intersection of the rhombus and the strip between L and M is a hexagon whose perimeter does not depend upon the choice of L, M.

Now let F be a figure of constant width h and K be a disk of the same width h. It is clear that the squares circumscribed about F and K are congruent and, consequently, have the same perimeter. We can use this fact as the basis for induction.

Assume that the 2^n-gons with equal angles circumscribed about F and about K have the same perimeter. Let us prove now that the same is true for 2^{n+1}-gons. Indeed, consider two adjacent sides $[b,e]$, $[b, f]$, and the opposite sides $[d,g]$, $[d,h]$ (Figure 15.9). The straight lines containing these sides form a parallelogram with altitudes equal to h, i.e., a rhombus. Similarly, we construct the rhombus for F. These rhombuses about K and F are congruent (Figure 15.9). Now we construct support lines L, M for K that are perpendicular to the diagonal $[b, d]$ and similar support lines L', M' for F. The distance between L and M (and between L' and M' as well) is equal to h. According to the above auxiliary proposition, the obtained polygons (that is, the intersections of 2^n-gons with the constructed strips) have equal perimeters. Carrying out this construction for each pair of adjacent sides, we obtain 2^{n+1}-gons with equal angles circumscribed about K and F and thus prove that these 2^{n+1}-gons have equal perimeters.

We showed that 2^n-gons with equal angles circumscribed about K and F have equal perimeters for every n. By increasing n without bound we can show that the boundaries of K and F have the same length. But for K this length is equal to πh. Consequently, the length of the boundary of F is equal to πh, too. ■

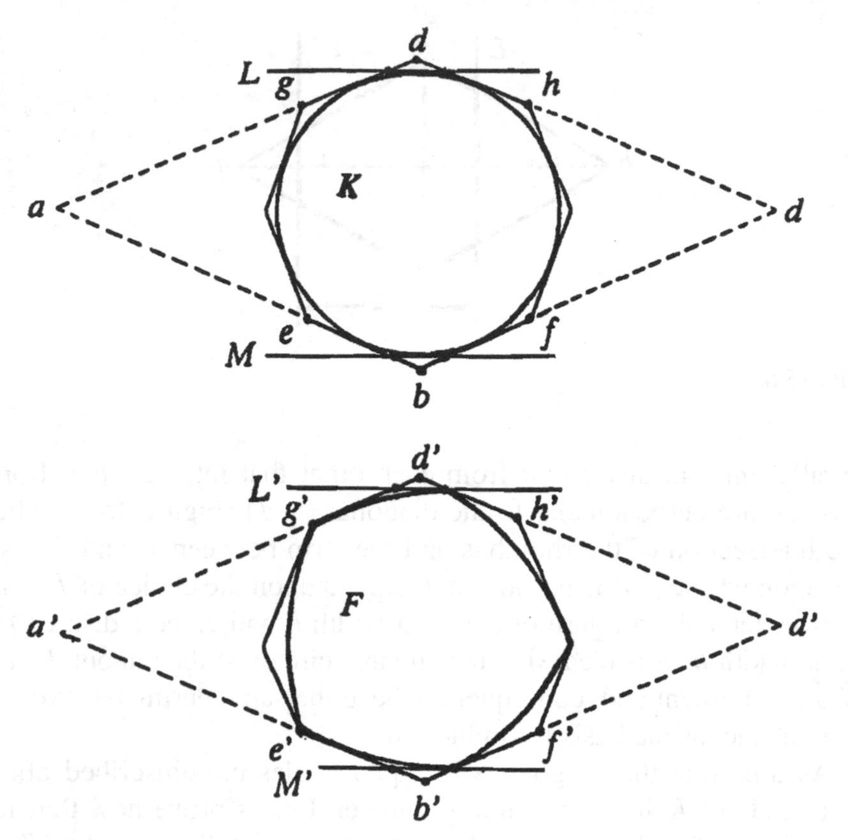

Fig. 15.9

In order to state the next theorem we introduce the operation *vector addition of convex figures*. Let F_1 and F_2 be convex figures in the plane, and assume that the origin point is chosen. Each point a of the plane is identified with the vector starting at the origin and ending at a (Figure 15.10).

By $F_1 + F_2$ we denote the set of points $a_1 + a_2$ where a_1 and a_2 belong to F_1 and F_2, respectively (Figure 15.11).

We leave it to the reader to prove the following properties of vector addition of convex figures.

For every pair of convex figures F_1 and F_2, the figure $F_1 + F_2$ is also convex.

If F_1 and F_2 are two nonparallel segments, then $F_1 + F_2$ is a parallelogram with sides congruent and parallel to F_1 and F_2 (Figure 15.12).

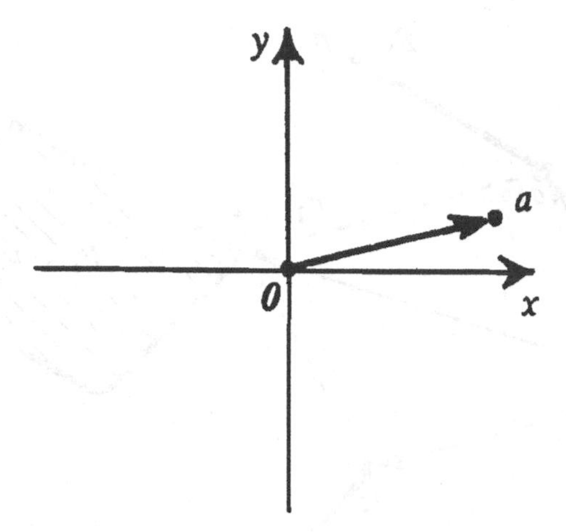

Fig. 15.10

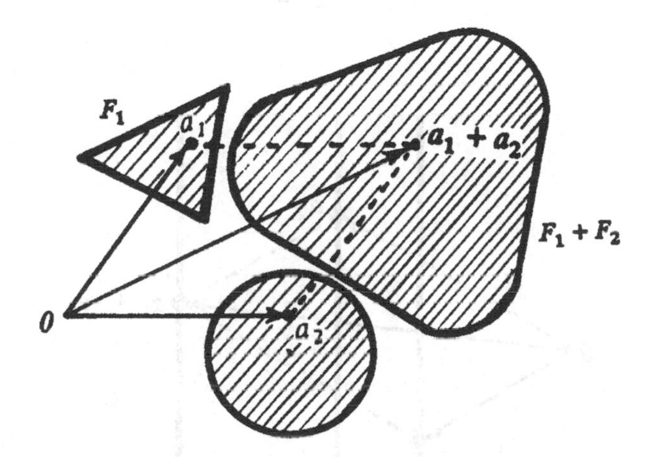

Fig. 15.11 Addition for figures

If L_1 and L_2 are parallel lines (segments), then $L_1 + L_2$ is a line (segment) that is parallel to L_1 and L_2 (Figure 15.13).

Finally, if L_1 and L_2 are parallel support lines of convex figures F_1 and F_2, and F_1 and F_2 are located on the same side of these lines (say "below"), then $L_1 + L_2$ is a support line of the convex figure $F_1 + F_2$ (Figure 15.14).

This implies that the width of the figure $F_1 + F_2$ in a certain direction is equal to the sum of the widths of the figures F_1 and F_2 in the same direction (Figure 15.15).

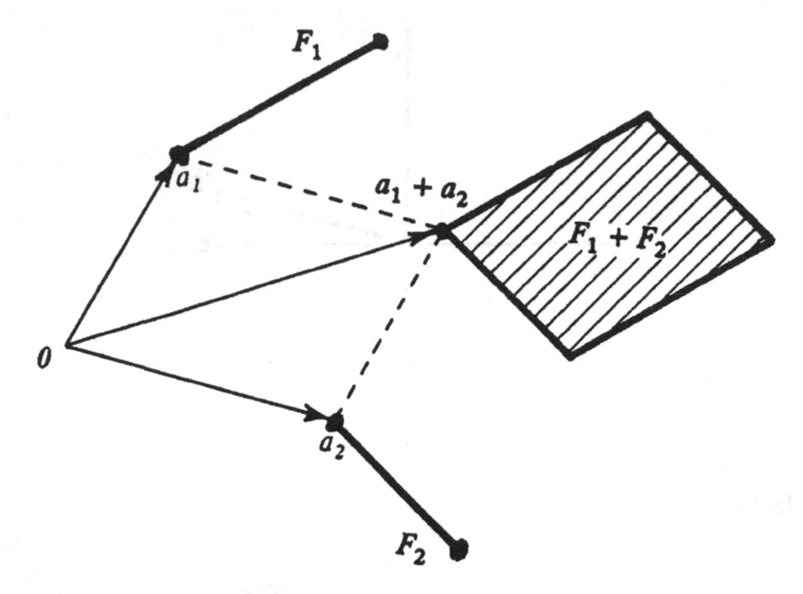

Fig. 15.12

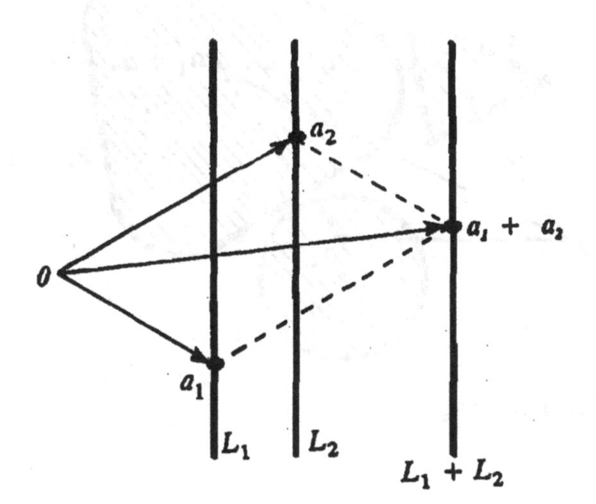

Fig. 15.13

Now we can prove the following proposition:

Theorem 15.2. *Let F be a plane figure and F' be the figure that is centrally symmetric to F with respect to the origin. The figure F has constant width h if and only if the vector sum $F + F'$ is the disk of radius h with center at the origin (Figure 15.16).*

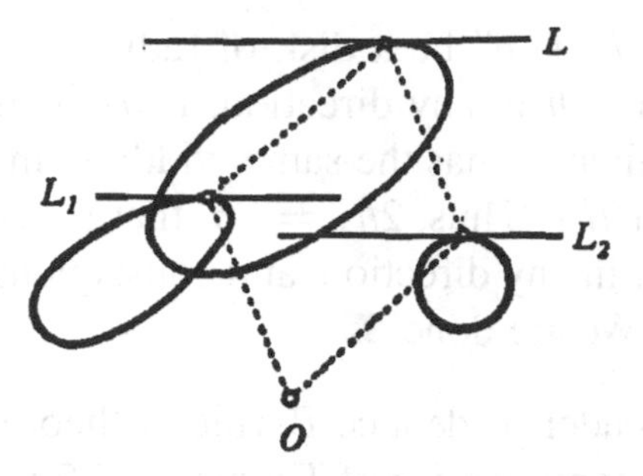

Fig. 15.14

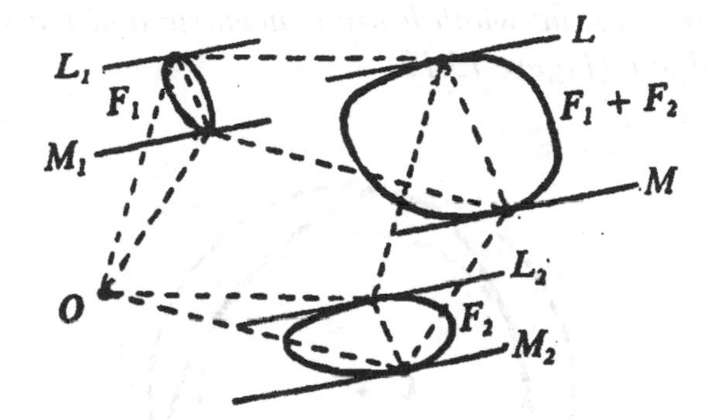

Fig. 15.15

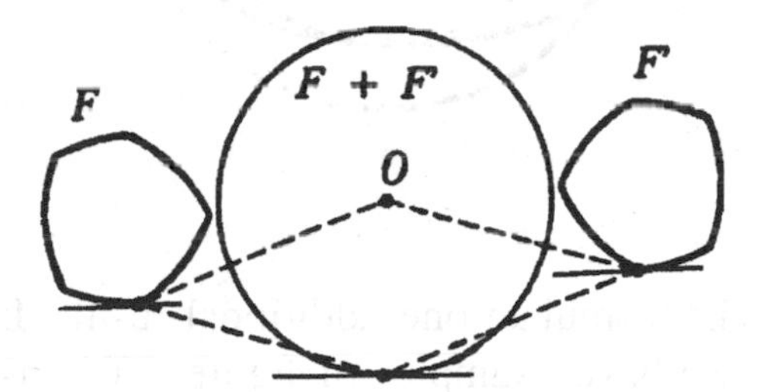

Fig. 15.16

Proof. Let F be a figure of constant width h. The width of the figure $F + F'$ in a certain direction is equal to the sum of the widths of the figures F_1 and F_2 in the same direction. Hence, the width of $F + F'$ in any direction is equal to $2h$. Moreover, $F + F'$ is centrally symmetric with respect to the origin. This means that each support line of $F + F'$ is distance h apart from the origin. Consequently, $F + F'$ is a disk of radius h.

Conversely, let $F + F'$ be a disk of radius h. Then the width of $F + F'$ is equal to $2h$ in any direction. If h_1 is the width of F in a certain direction, then F' has the same width h_1 in this direction, and $F + F'$ has width $2h_1$. Thus, $2h_1 = 2h$, that is, $h_1 = h$. This means that F has width h in any direction, and consequently, F is a figure of constant width h. We are done. ■

We invite the reader to deduce Barbier's theorem from the above theorem. Another consequence of Theorem 15.2 is the following:

Theorem 15.3. *The inscribed and circumscribed circles of an arbitrary figure of constant width h are concentric and the sum of their radii is equal to h (Figure 15.17).*

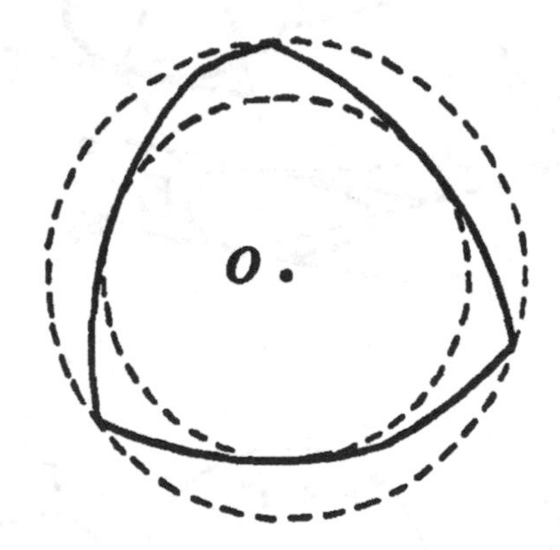

Fig. 15.17

Finally, we will formulate one additional result. In Figures 15.3, 15.4, and 15.5, we have examples of figures of constant width that are formed by several arcs of circles. In general, such figures can be constructed by the following method. We take a polygon with an odd number of vertices that has the following property: for every vertex there are two diagonals of length h emanating from that vertex, and all other diagonals are shorter than h (Figure 15.18). Now we construct all disks of radius h with centers at the vertices of this polygon. Then the intersection of all these disks is a figure of constant width h (Figure 15.19) that is bounded by several arcs of radius h.

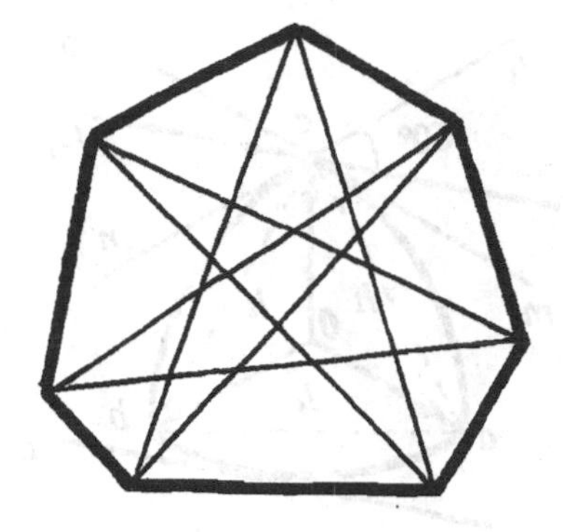

Fig. 15.18

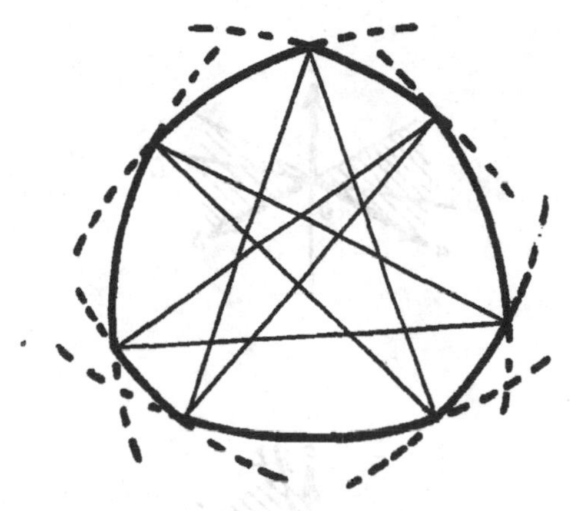

Fig. 15.19

Our result is this:

Theorem 15.4. *Every figure of constant width h can be represented as the limit of a convergent sequence $M_1, M_2, \ldots$ where each M_k is a figure of constant width h that is bounded by several arcs of radius h (i.e., each M_k is constructed as shown in Figure 15.19.)*

Solutions to Exercises

15.1. The tangent line *mc* to the arc *ac* at the point *c* (Figure 15.20) is perpendicular to the chord $[c,b]$. Consequently, the angle *mca* is 30°, and the angle *mcn* is 120°. Let L_1 and L_2 be two parallel support lines of the Reuleaux triangle F. Let us construct the line L parallel to L_1 and L_2 through the origin. Then three cases are possible: Figures 15.21, 15.22, and 15.23. In the first case (Figure 15.21) one of the lines

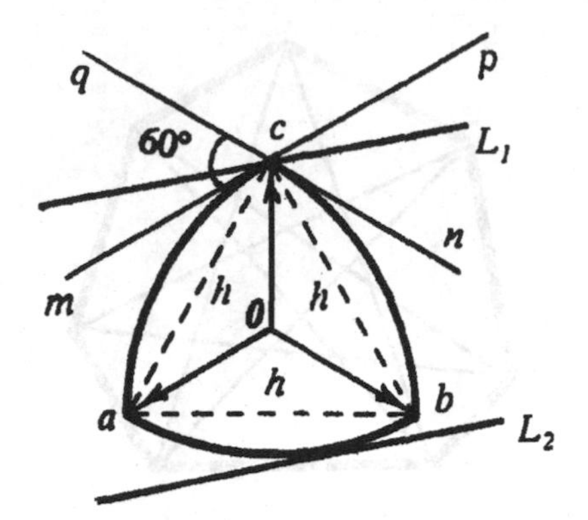

Fig. 15.20

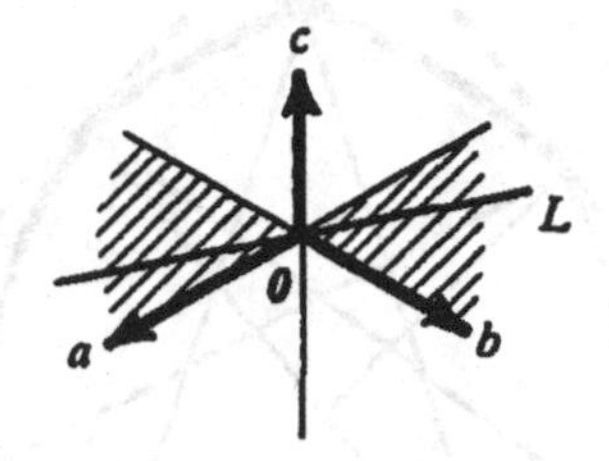

Fig. 15.21

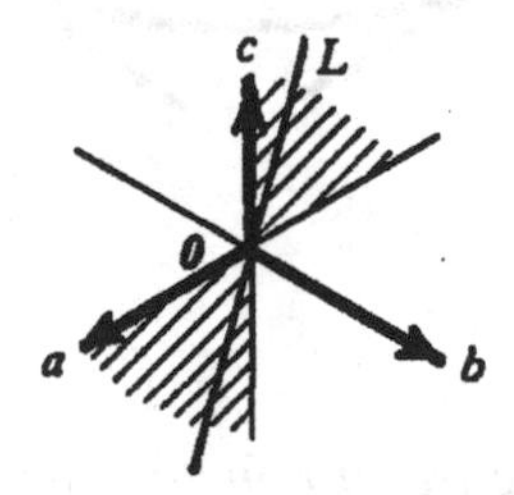

Fig. 15.22

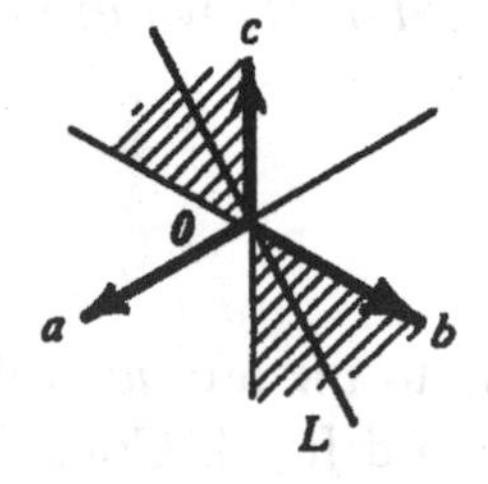

Fig. 15.23

L_1, L_2 (say L_1) passes through c, and the other line (L_2) is tangent to arc ab. Hence, the distance between two support lines is equal to the radius of ab, i.e., it is equal to h. The other two cases (Figures 15.22 and 15.23) are similar. ■

15.2. The solution is similar to the solution to Exercise 15.1 above. If one of two parallel support lines passes through a vertex of M, then the other line is tangent to the opposite arc (Figure 15.24). Therefore, the distance between two parallel support lines is equal to the length of the longest diagonal of M. ■

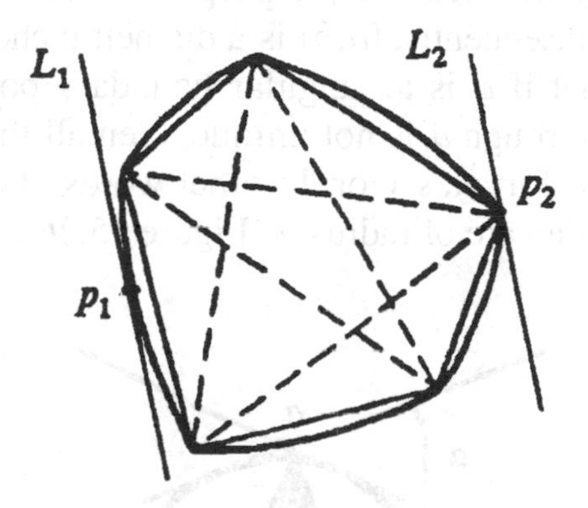

Fig. 15.24

15.3. Let a, b be two points of a figure F of constant width h. Further, let L_1, L_2 be two support lines of F that are perpendicular to $[ab]$ with the corresponding support points p_1, p_2, and p_1q parallel to ab. Then $|ab| \leq |p_1q| = h$ (Figure 15.25). This is true for any pair of points a, b of F, and consequently, $d \leq h$.

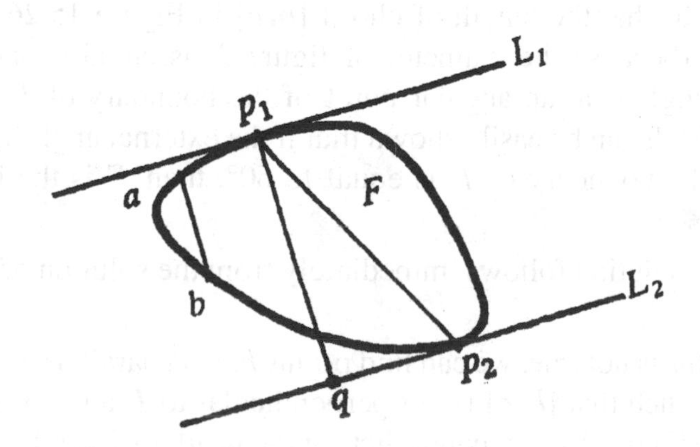

Fig. 15.25

Conversely, let L_1, L_2 be two parallel support lines of F, and p_1, p_2 be the corresponding support points. Then $|p_1p_2| \geq |p_1p_2| = h$ (Figure 15.25). Hence, $d \geq h$.

This reasoning shows that q must coincide with p_2 in Figure 15.25. In other words, if L_1, L_2 are two parallel support lines of a figure of constant width and p_1, p_2 are the corresponding support points, then the segment $[p_1, p_2]$ is perpendicular to L_1 and L_2. ■

15.4. This follows immediately from the solution to Exercise 15.3. Indeed, let L be a support line through a and M the support line parallel to L through b. Then $[a,b]$ is perpendicular to L and $|ab| = h$ (Figure 15.7). Consequently, $[a,b]$ is a diametral chord of F.

Let us note that if a is an angular boundary point of F, that is, the support line through a is not unique, then all the diameters of F emanating from a form a sector. In other words, the boundary of the figure F contains an arc of radius h (Figure 15.26).

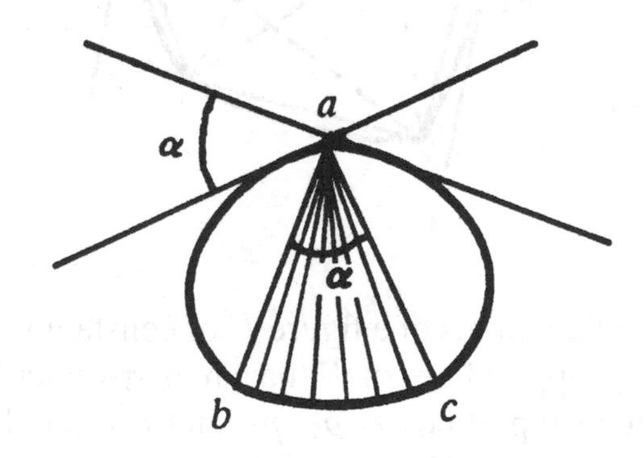

Fig. 15.26

It is clear that the length of chord $[b, c]$ in Figure 15.26 does not exceed h (because the diameter of figure F is equal to h). So, the external angle α at an angular point of the boundary of F does not exceed 60°. It can be easily shown that if the external angle at a corner point of the boundary of F is equal to 60°, then F is the Reuleaux triangle. ■

15.5. The solution follows immediately from the solution of Exercise 15.3. ■

15.6. If this is not true, we can find points b, c on parallel support lines L and M such that $[b, c]$ is not perpendicular to L and consequently $|bc| > h$ (Figure 15.27), contradicting the result of Exercise 15.3. ■

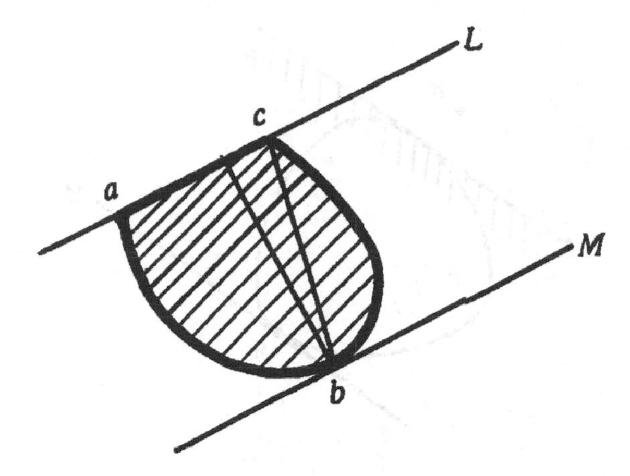

Fig. 15.27

15.7. Indeed, if two diametral chords $[a, d]$ and $[b, c]$ have no common points, then we obtain a convex quadrangle $abcd$ inscribed in F with two opposite sides of length h (Figure 15.28). But then at least one of the diagonals of this quadrangle has length greater than h (can you explain why?), contradicting the result of Exercise 15.3. ■

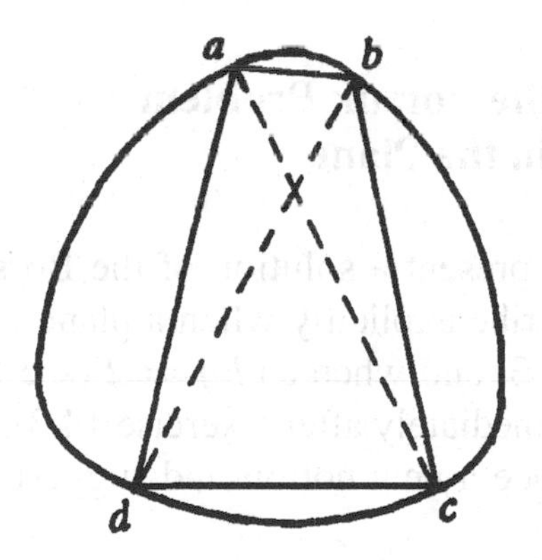

Fig. 15.28

15.8. Suppose that Q does not coincide with F, that is, there is a point a in Q that does not belong to F. Then there exists a support line L of the figure F such that F is located in a closed half-plane P with boundary L, but a does not belong to P (Figure 15.29).

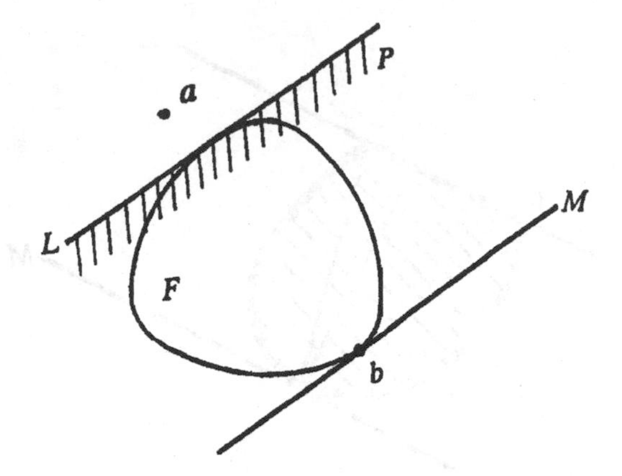

Fig. 15.29

Let us denote the support line of F parallel to L by M, and the corresponding support point by b. Then $|ab|$ is greater than the width of the strip between L and M, that is, $|ab| > h$. But both a and b belong to Q, contradicting the assumption that the diameter of Q is equal to h. This contradiction shows that Q coincides with F. ■

16 Solution of the Borsuk Problem for Figures in the Plane

In this section, we present a solution of the Borsuk problem for the plane, i.e., we describe explicitly when a plane figure F satisfies the condition $\alpha(F) = 3$, and when $\alpha(F) = 2$ (see the definition of the Borsuk number immediately after Exercise 14.3).

First, we introduce a new notion and suggest several exercises to the reader.

Let F be a plane figure of diameter not exceeding h. We call intersection of all disks of radius h containing F (Figure 16.1) the *h-hull* of F and denote it F^*.

Exercise 16.1. Let F be a plane figure of diameter not exceeding h. Prove that F and its h-hull F^* have the same diameter.

Exercise 16.2. Let F be a figure of constant width h. Prove that its h-hull F^* coincides with F.

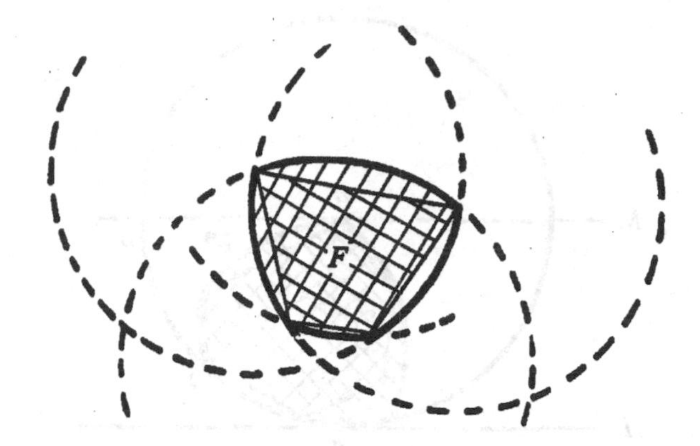

Fig. 16.1

Exercise 16.3. Let F be a figure of constant width h, L its support line, and a the corresponding support point. Further, let K be the disk of radius h tangent to L at a and situated in the half-plane that contains F (Figure 16.2). Prove that F is contained in K.

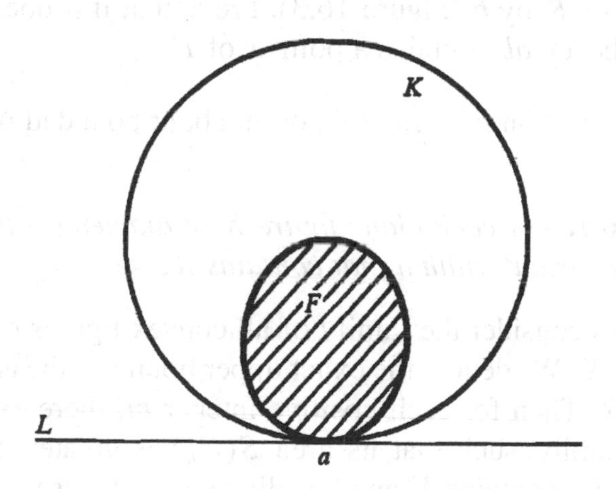

Fig. 16.2

Exercise 16.4. Let F be a plane figure of diameter h and F^* its h-hull. Prove that if the width of F^* in at least one direction is less than n, then $\alpha(F) = 2$.

Exercise 16.5. Let F be a plane figure of diameter h such that its h-hull F^* is a figure of constant width h. Let L be a support line for F^* with corresponding support point a and K be the disk as in Exercise 16.3 (Figure 16.2).

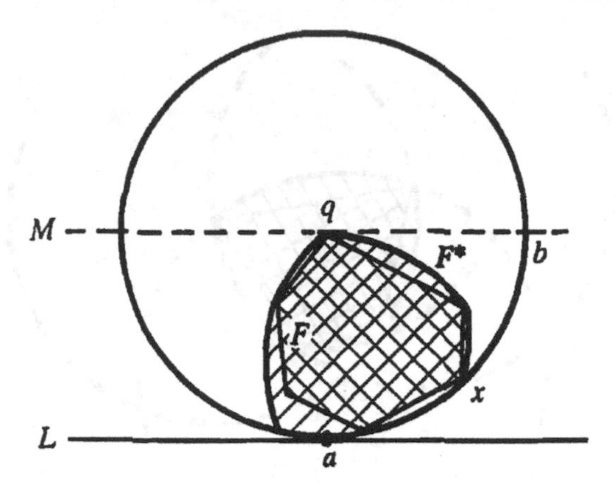

Fig. 16.3

The center q of K is the support point corresponding to the support line M for F^* parallel to L (can you see how this follows from Exercises 15.4 and 15.5?). Denote the intersection of M and the circumference of K by b (Figure 16.3). Prove that if a does not belong to F, then the arc ab contains a point x of F.

Now we prove an important theorem about bounded figures in the plane.

Theorem 16.1. *For each plane figure X of diameter h there exists a figure F of constant width h that contains X.*

Proof. Let us consider the family of all convex figures of diameter h that contain X. We denote the exact upper bound of the areas of these figures by R. Then for each positive integer m, there exists a figure F_m in the family, such that its area $S(F_m)$ is greater than $R - \frac{1}{m}$. Each figure F_m contains X and has diameter h. Consequently, all the figures F_m are situated in the disk of radius h with center at fixed point c of the figure X. By Blaschkes Theorem 13.2, the sequence F_1, $F_2, \ldots, F_n, \ldots$ contains a converging subsequence. Let F be the limit of this subsequence. It is clear that $S(F) = R$; that is, F has maximal area among figures of diameter h containing X.

Let us prove that F is a figure of constant width h. Evidently the h-hull F^* of F coincides with F (otherwise, F^* would be a figure of greater area with diameter h containing X, which is impossible). Assume that $F = F^*$ is not a figure of constant width h. Then

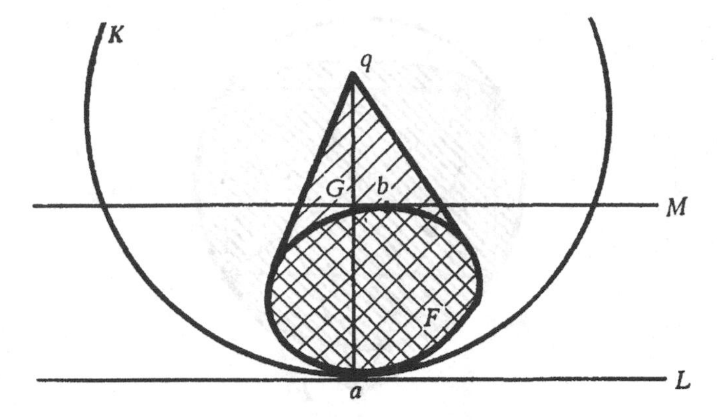

Fig. 16.4

there exists two parallel support lines L, M of F with distance between them less than h. Let a, b be the corresponding support points (Figure 16.4), K be the disk of radius h with center q as in Exercise 16.3. Then q does not belong to F. Taking the convex hull of the union of the point q and the figure F, we obtain a convex figure G of diameter h that contains F and has a greater area, contradicting the choice of F. This contradiction proves that F is a figure of constant width h. This completes the proof. ■

Exercise 16.6. Let F be a plane figure of diameter h. Prove that if its h-hull F^* is not a figure of constant width, then there exists at least two distinct figures of constant width h each containing F.

Finally, let us consider a complete solution to Borsuk's problem in the plane (found in [B2]). According to Theorem 16.1, each figure X of diameter h is contained in a figure $F$of constant width h. Is this figure F unique for the given X? Simple examples show that the answer is sometimes "yes" and sometimes "no." For example, if X is an equilateral triangle with side length h, then there exists only one figure of constant width h containing X, namely, the Reuleaux triangle. We have encountered other examples of this in Exercise 15.2. Let now X be a segment of a disk of diameter h (i.e., the intersection of the disk and a half-plane) that is larger than the half-disk (Figure 16.5). Then there are many different figures of constant width h containing X.

One of them is the disk of diameter h. The second one is shown in Figure 16.5 (the dotted segments have length h). Now we are ready to prove the main result.

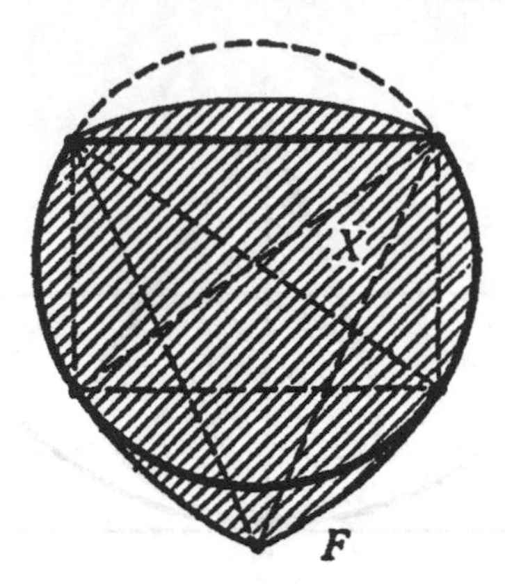

Fig. 16.5

Boltyanski's Theorem 16.2. (V. Boltyanski [B2]) Let X be a figure in the plane of diameter h. Then $a(X) = 3$ if and only if the figure of constant width h containing X is unique.

Proof. Assume that there exists only one figure F of constant width h that contains X. We must prove that $a(X) = 3$. Assume the opposite, i.e., $a(X) = 2$. This means that X is a union of two figures Q_1 and Q_2, each of which has diameter less than h. Consider the h-hulls X^*, Q_1^*, Q_2^* of the figures X, Q_1, Q_2, respectively. According to Exercise 16.6, X^* is a figure of constant width h, that is, X^* coincides with F. Let us prove that the boundary of F is contained in the union of figures Q_1^* and Q_2^*.

Indeed, let a be a boundary point of the figure F. We construct a support line L of F through a and an other support line M of F parallel to L (Figure 16.6) and denote the corresponding support point by b. Then due to Exercise 15.5 the segment $[a,b]$ is perpendicular to L and M. We have two possibilities: i) a is in X; ii) a is not in X.

In case (i) the point a belongs to Q_1 or Q_2, and consequently, a belongs to Q_1^* or Q_2^*.

Let us consider case (ii). According to Exercise 16.5, the arc ap in Figure 16.6 contains a point x *of* X. Similarly, the arc aq contains a point y of X. Now, if b does not belong to X, then, clearly, the arc br contains a point z of the figure F. But the distance between y and z is greater than h (can you explain why?), contradicting the assumption that X has diameter h. Consequently, b belongs to X, that is, b is contained in Q_1 or Q_2 (say in Q_1). Further, since both $|bx|$ and $|by|$

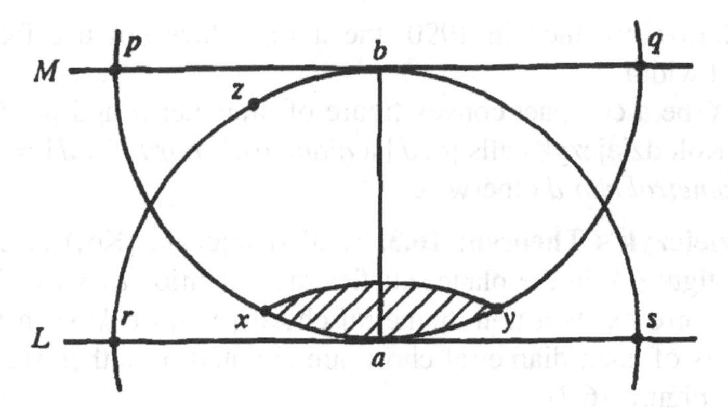

Fig. 16.6

are equal to h, both x and y belong to Q_2 (because the diameter of Q_1 is smaller than h). But the lens bounded by two arcs of radius h passing through x and y (shaded in Figure 16.6) is contained in *each* disk of radius h that includes x and y. So, Q_2^*, which is the intersection of all disks of radius h containing Q_2, contains this lens. This means that a belongs to Q_2^*.

Thus, in both cases (i) and (ii) the point a of bdF is contained in the union of the figures Q_1^* and Q_2^*. In other words, bdF can be decomposed into two figures of diameters smaller than h (due to Exercise 16.1, the diameters of Q_1^* and Q_2^* are equal to the diameters of Q_1 and Q_2, respectively). But this is impossible (this can be proven by the same method, as in the solution to Exercise 14.1). This contradiction shows that the equality $a(X) = 2$ is impossible, and therefore $a(X) = 3$.

Conversely, assume that there exist at least two distinct figures F_1, F_2 of constant width h that contain X. Then the intersection ϕ of F_1 and F_2 is not a figure of constant width h.

Moreover, ϕ contains X, and ϕ is the intersection of a family of disks of radius h containing X (since from Exercise 16.3, we know F_1, and F_2 are the intersections of some family of such disks). So, X^* is contained in ϕ, and consequently, X^* is not a figure of constant width h. Therefore, by Exercise 16.4, we have $a(X) = 2$. This completes the proof. ■

We would like to bring to your attention another solution of the Borsuk problem for plane figures that was found by the Polish mathematician Krzysztof Kolodziejczyk [Ko]. In this very interesting solution obtained in 1990, the author does not use figures of constant width.

Let X be a compact convex figure of diameter h and $[c, d]$ be its chord. Kolodziejczyk calls $[c, d]$ a *diametral chord* if $|cd| = h$ and a *non-diametral chord* otherwise.

Kolodziejczyk's Theorem 16.3. (Kolodziejczyk, [Ko]) A compact convex figure X in the plane satisfies the condition $a(X) = 2$ if and only if there exists a non-diametral chord $[c, d]$ of X such that the endpoints of each diametral chord are situated on either side of the line cd (Figure 16.7).

For example, the equilateral triangle T whose side has length h does not satisfy the indicated condition. Indeed, for every non-diametral chord $[c, d]$ we can find a diametral chord situated in one closed half-plane with the boundary line cd (Figures 16.8 and 16.9).

Exercise 16.7. Using Theorem 16.2, prove Kolodziejczyk's Theorem 16.3.

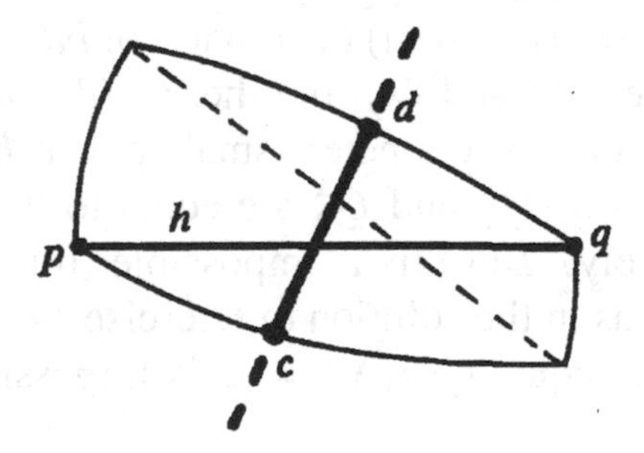

Fig. 16.7

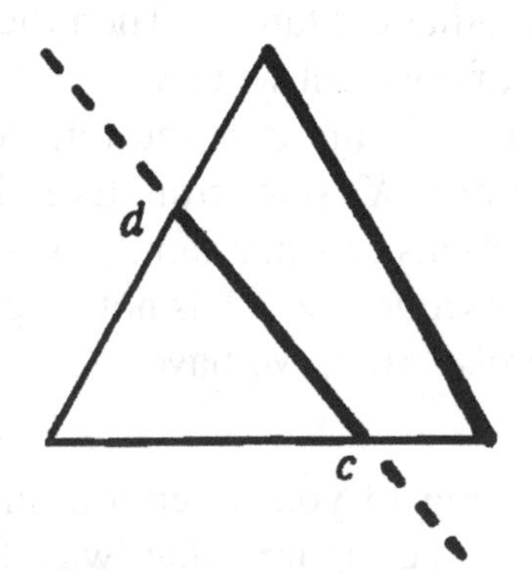

Fig. 16.8

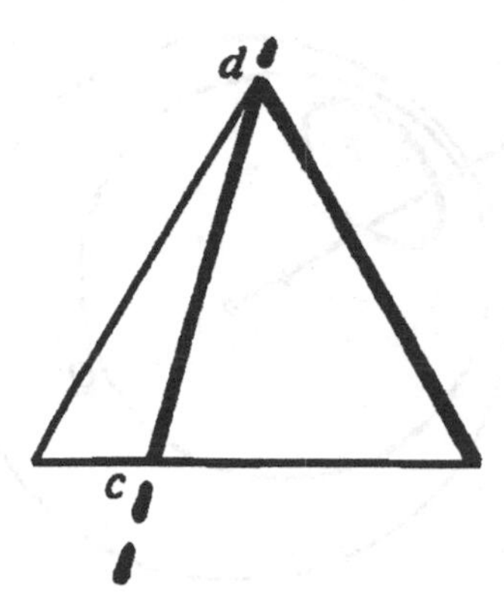

Fig. 16.9

Solutions to Exercises

16.1. Assume at first that the diameter d of F is equal to h. Let x be a point of F^* and a a point of F. The circle K_a with center a and radius d contains the whole figure F (since $|ap| \leq d$ for every point p of F). Thus, K_a is one of the disks that has the figure F^* as its intersection. This means that F^* is contained in K_a, and consequently, $|ax| \leq h$. This inequality is truc for each point a in F. Hence, F is contained in the disk K_x with center x and radius h. So, K_x has F^* as its intersection. Therefore, F^* is contained in K_x. Thus, $|xy| \leq h$ for every point y of F^*, and consequently, the diameter d_1 of F^* does not exceed h. On the other hand, it is clear that $d_1 \geq h$ since F^* contains F. This completes the proof in this case.

Let us assume now that the diameter d of F is less than h. We denote by F^{**} the d-hull of F. In view of the above reasoning, the diameter of F^{**} is equal to d. It can be easily shown that the h-hull F^* of F is contained in F^{**}. Indeed, if a point z is not in F^{**}, then there exists a disk Q of radius d such that Q contains F, and z is not in Q. Hence, there exists a disk K of radius h that contains Q and does not contain z (since $d < h$; see Figure 16.10). Then K is one of the disks that have F^* as their intersection, and consequently F^* is contained in K. This means that z is not in F^*.

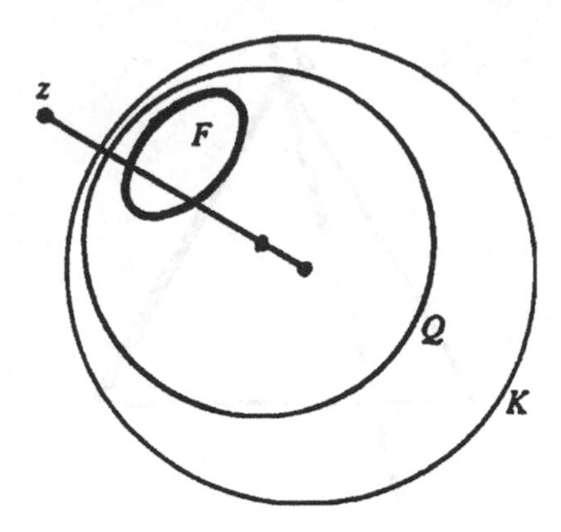

Fig. 16.10

Thus, if z is not contained in F^{**} then z is not in F^*, that is, F^* is contained in F^{**}. Hence the diameter d_1 of F^* does not exceed d. On the other hand, $d_1 \geq d$ (since F^* contains F). This completes the proof. ■

16.2. The diameter of the figure F is equal to h, and consequently, the diameter of F^* is equal to h as well (Exercise 16.1). The validity of the assertion of Exercise 16.2 follows immediately from Exercise 15.8 of the previous section. ■

16.3. Let us prove a more general proposition from which Exercise 16.3 follows by virtue of Exercise 16.2:

Let F be a figure of diameter h and F^* its h-hull. Further, let L be a support line of F^*, a the corresponding support point, and K the disk of radius h that is tangent to L at a and situated in the half-plane P that contains F^*. Then F^* is contained in K.

Indeed, assume that there exists a point x in F^* that does not belong to K (Figure 16.11). Denote by M the "lens" bounded by two arcs of radius h passing through the points a and x. Each disk of radius h containing a and x contains the whole lens M. Consequently, this lens is contained in F^* (since both points a and x belong to F^*, and F^* is the intersection of a family of disks of radius h). But M is not contained in the half-plane P, contradicting the assumption that L is a support line of F^*. This contradiction proves the desired statement. ■

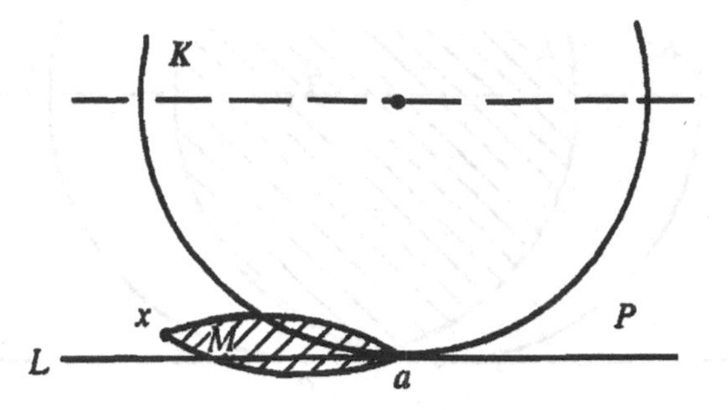

Fig. 16.11

16.4. Let L and M be two parallel support lines of F^* such that the distance between them is less than h (Figure 16.12), and a and b be the corresponding support points. Then F^* is contained in the disk K of radius h tangent to L at a. Similarly, F^* is contained in the disk Q of radius h tangent to M at b. The intersection of K and Q is a lens bounded by two arcs of radius h. The centers of K and Q are situated outside the lens, and consequently, the distance between the endpoints p and q of the lens is less than $2h$. Therefore, the line passing through the centers of the disks decomposes the lens (and the figure F^* contained in the lens) into two parts of diameter less than h. ■

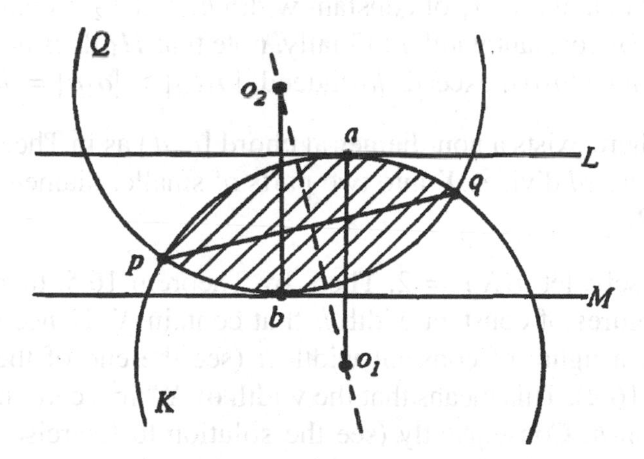

Fig. 16.12

16.5. If the arc ab does not contain any point of F, then for small enough $\mu > 0$ the image K^1 of K under the translation through the vector $\mu\overrightarrow{bq}$ contains F (Figure 16.13). Consequently, K^1 is one of

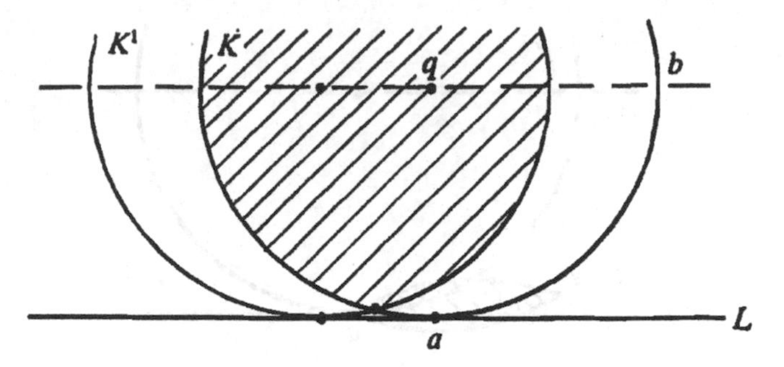

Fig. 16.13

the disks that have F as their intersection, and hence F^* is contained in K^1. Therefore, F^* is contained in the intersection of K and K^1, contradicting the assumption that L is a support line of F^*. ■

16.6. Let us repeat the construction discussed in the solution to Exercise 16.4 (see Figure 16.12). The convex hull of the union F_1 of the figure F and the center o_1 of the disk K is a figure of diameter h (indeed, the distance between o_1 and any point x of the figure F does not exceed h since F is contained in the lens). Similarly, the convex hull of the union F_2 of the figure F and the center o_2 of the disk Q is a figure of diameter h. Consequently, by Theorem 16.1, F_1 is contained in a figure H_1 of constant width h, and F_2 is contained in a figure H_2 of constant width h. Finally, note that H_1 does not coincide with H_2 since $|o_1o_2|$ exceeds h (indeed, $|o_1o_2| > |o_1a| = h$). ■

16.7. If there exists a non-diametral chord $[c, d]$ as in Theorem 16.3, then the line cd divides X into two parts of smaller diameters, that is, $a(X) = 2$.

Conversely, let $a(X) = 2$. Then, by Theorem 16.5, there are two distinct figures of constant width h that contain X. Hence the h-hull X^* is not a figure of constant width h (see the end of the proof of Theorem 16.2). This means that the width of X^* in a certain direction is less than h. Consequently (see the solution to Exercise 16.4), the h-hull X^* is contained in a lens with anglular points p and q such that $|pq| < h\sqrt{3}$ (see the notation in Figure 16.12). This means that the endpoints of each diametral chord of X^* (and of X, too) are situated on either side of the line o_1o_2. Thus, o_1o_2 intersects X in the required non diametral chord $[c, d]$. ■

17 Illumination of Convex Figures

In Figure 17.1 you see a convex figure F. A parallel beam of light, defined by a vector $\vec{w}$, illuminates part of its boundary. The points a and b are not illuminated. In other words, a boundary point c is *illuminated*, if there is $\mu > 0$ such that the point c^1, $\overrightarrow{cc^1} = \mu\vec{w}$ is in the interior of F.

The *illumination problem* asks for the minimal number of parallel beams that illuminate the entire boundary of a given convex figure F. We denote this minimal number by $c(F)$. Figure 17.2 shows that for a disk F we have $c(F) = 3$. And if F is a parallelogram, then

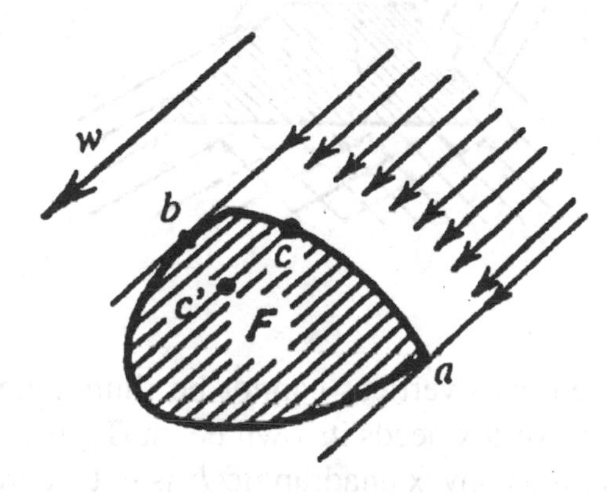

Fig. 17.1

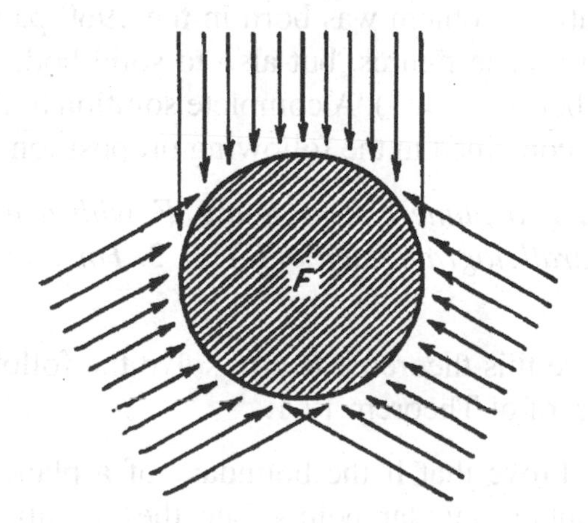

Fig. 17.2

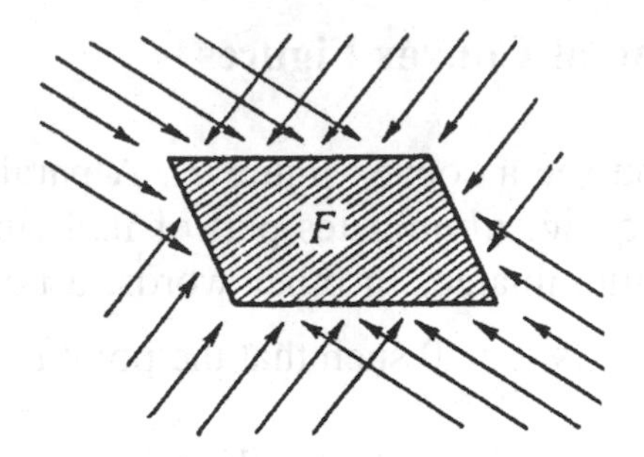

Fig. 17.3

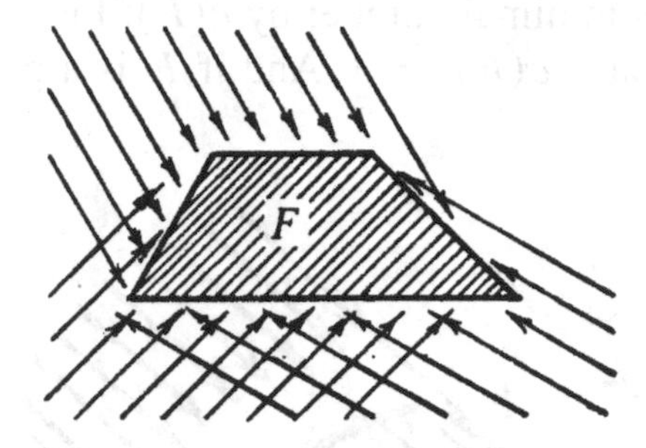

Fig. 17.4

$c(F) = 4$. Indeed, two vertices cannot be illuminated by the same beam, that is, each vertex needs its own beam (Figure 17.3). It can be easily shown that if a convex quadrangle F is not a parallelogram (for example, if F is a trapezoid), then $c(F) = 3$ (Figure 17.4).

The illumination problem was born in the 1960 paper [B3]. It applies not only to plane figures, but also to solid bodies (and even to n-dimensional bodies, $n > 3$). A complete solution to the problem for plane figures is contained in the following proposition:

Theorem 17.1. *If a plane convex figure F with a non-empty interior is not a parallelogram, then $c(F) = 3$. For any parallelogram $P, c(P) = 4$.*

Can you prove this theorem? If not, solve the following exercises that lead to a proof of Theorem 17.1.

Exercise 17.1. Prove that if the boundary of a plane convex figure F does not contain angular points (see the definition right before Exercise 13.3), that is, all the boundary points of F are regular, then $c(F) = 3$.

Exercise 17.2. Let F be a plane convex figure and p an angular point on its boundary. Prove that there exists a parallelogram *pqrs* circumscribed about F such that the rays emanating from p and passing through q and s are tangent to F (Figure 17.5).

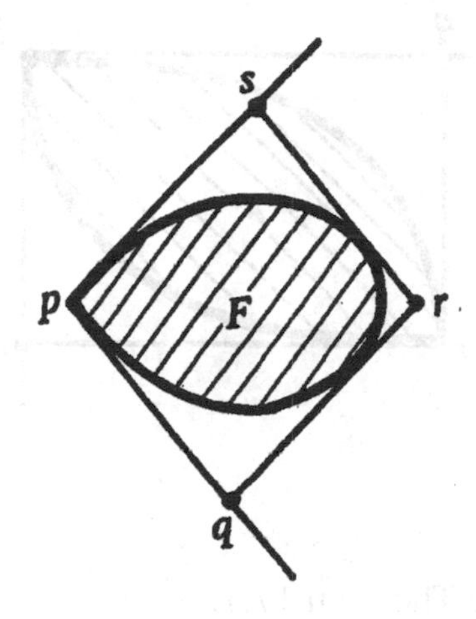

Fig. 17.5

Exercise 17.3. Let F be a plane convex figure, p an angular point on its boundary and *pqrs* the circumscribed parallelogram, as in Exercise 17.2. Prove that if the point r does not belong to F (Figure 17.6), then $c(F) = 3$.

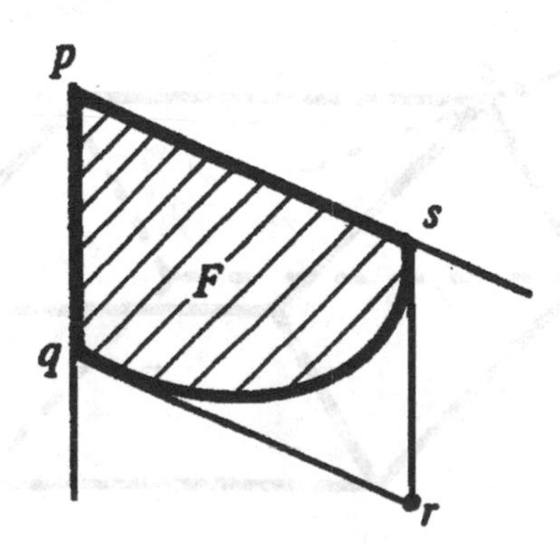

Fig. 17.6

Exercise 17.4. Let F be a plane convex figure, p an angular point on its boundary and $pqrs$ the circumscribed parallelogram, as in Exercise 17.2. Prove that if at least one of the points q, s does not belong to F (Figure 17.7), then $c(F) = 3$.

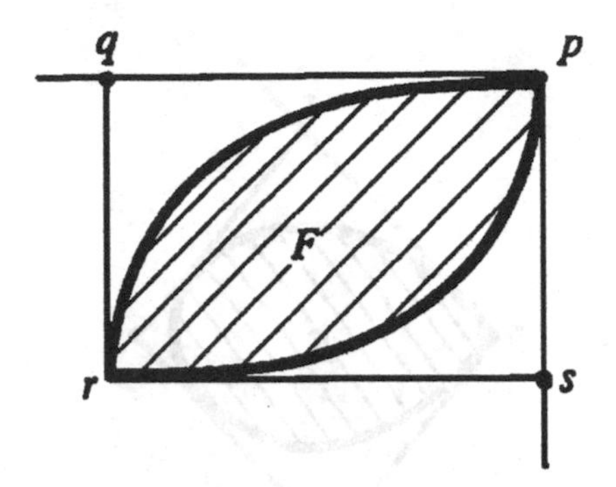

Fig. 17.7

Exercise 17.5. Prove Theorem 17.1.

We are done with Flatland. How are things in space? It is clear that if F is a parallelepiped, then $c(F) = 8$. Indeed, no two vertices of F can be illuminated by one light beam, that is, each vertex needs its own beam (Figure 17.8). Further, in the plane each a parallelogram was a figure of *maximal* value of $c(F)$. Does a similar proposition hold for the 3-space? In other words,

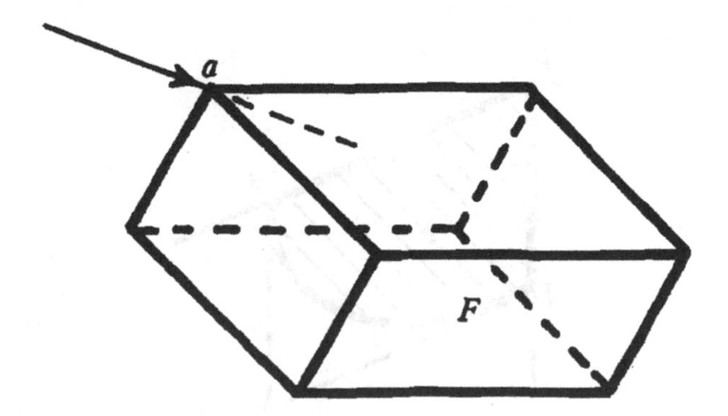

Fig. 17.8

Problem 17.1. *Is it true that for all 3-dimensional convex bodies F, except parallelepipeds, the inequality $c(F) < 8$ holds?*

This is still an open problem! Even for polyhedrons nobody knows the answer. The answer is known only for centrally symmetric convex bodies; the Polish mathematician M. Lassak proved a positive answer for them. In n-dimensional space R^n this question is known as the *Hadwiger Problem:*

Hadwiger's Illumination Open Problem 17.2. *Is it true that for all convex bodies F in R^n, except parallelotopes (i.e., n-dimensional analogues of parallelepipeds), the inequality $c(F) < 2^n$ holds?*

When $n > 4$ the answer is unknown even for centrally symmetric polytopes. But in several particular cases a solution to the Hadwiger problem is known.

Exercise 17.6. Prove that if a convex body F in R^3 is smooth (that is, each boundary point of F is regular), then $c(F) = 4$.

A similar result was proven in 1945 by the Swiss mathematician Hugo Hadwiger [H1] for n-dimensional spaces R^n for all positive integers n: $c(F) = n + 1$ for all smooth convex bodies F in R^n.

Moreover, if a convex body F in R^n has at most n angular points, then $c(F) = n + 1$ (see [B2]).

In R^3, a stronger result is known (see [K]): if a convex body F in R^3 has no more than 4 angular points, then $c(F) = 4$.

There is a very interesting relationship between the illumination problem and another combinatorial problem posed by Hadwiger in [H3].

Let F be a figure. Choose an arbitrary point q in the plane and a positive number k. For any point p of the figure F we can find a point p' on the ray qp such that $\frac{|qp'|}{|qp|} = k$ (Figure 17.9). The set of all points so obtained comprises a new figure F'. The transition from F to is F' called a *homothety* with center q and coefficient k, and F' itself is called a *homothetic image* of F. (Homothety with a negative coefficient will not be necessary for us in what follows, and we shall therefore not consider it.)

If a figure F is convex, then its homothetic image F' is also convex (can you prove it?).

We will call a figure F_1 a *smaller copy* of F if F_1 is homothetic to F with a center q and a positive ratio $k_1 < 1$ (Figure 17.9).

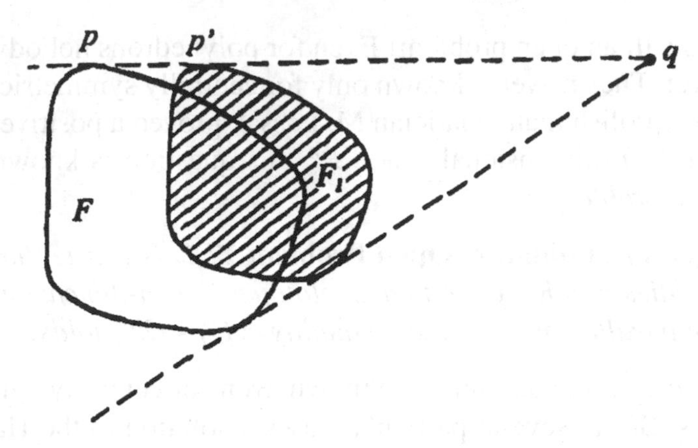

Fig. 17.9

The Hadwiger Covering Problem 17.3. *What is the minimal number n such that n smaller copies of a convex figure F can cover the entire figure F?*

This problem can be considered in the plane as well as in 3-space (and in n-dimensional space R^n). We denote the minimal number of smaller copies covering F by $b(F)$. Hadwiger found a solution to the problem for plane figures [H1]. He also proved that

for smooth convex bodies F; in R^n, $b(F) = n + 1$.

For example, if F is a disk, then $b(F) = 3$ (Figure 17.10). If F is a parallelogram, then $b(F) = 4$. Indeed, no two vertices of a parallel-

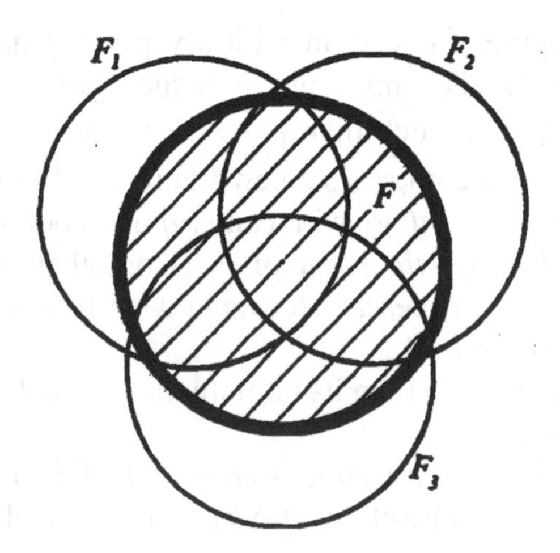

Fig. 17.10

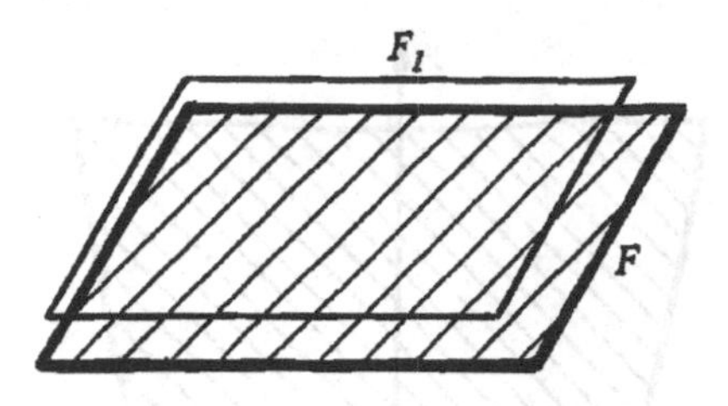

Fig. 17.11

ogram can be covered by its smaller copy (Figure 17.11). Thus, each vertex needs its own smaller copy.

The events in the illumination problem were very similar, weren't they? Perhaps these two problems have the same answer for every convex figure F. The following theorem explains when this is indeed true.

Theorem 17.2. *(V. G. Boltyanski, [B1]) For every compact convex body F in R^n, $b(F) = c(F)$.*

In particular, we have the following result for compact convex figures in the plane:

Exercise 17.7. Prove that for every convex plane figure F that is not a parallelogram, $b(F) = 3$.

The next exercise shows that Theorem 17.2 is in general false for unbounded figures.

Exercise 17.8. Let the figure F be defined by $y \geq x^2$ in the Cartesian coordinate system, that is, F is the convex figure whose boundary coincides with the parabola $y = x^2$. Figure 17.12 shows that $c(F) = 1$. Prove that $b(F) = \infty$.

Finally, let us look into a connection between the problems discussed in this section and the Borsuk problem.

Theorem 17.3. *For every compact convex body F in R^n, $a(F) \leq b(F)$. Thus, we have*

$$a(F) \leq b(F) = c(F).$$

This assertion is trivial. Indeed, if $b(F) = s$, than the body F can be covered by s smaller copies of F. And it is clear that if the diameter

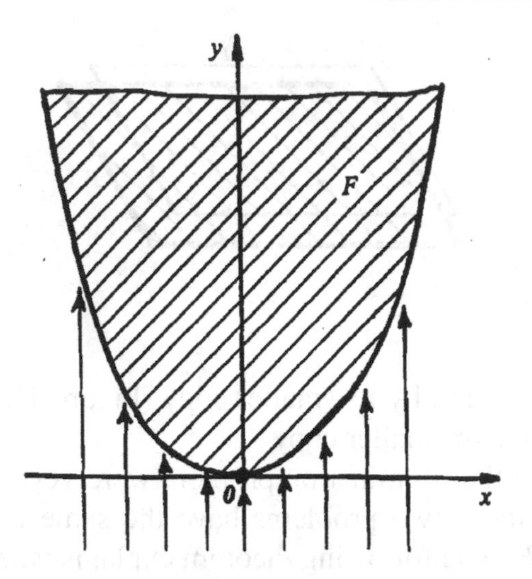

Fig. 17.12

of F is equal to d, then the diameter of each of its smaller copies is less that d. This completes the proof.

Finally, due to Exercise 17.6, if a convex body F in R^3 is smooth, then $a(F) \leq 4$, in agreement with the Eggleston result of Section 14.

A similar result holds in R^n: for every smooth convex body F in R^n, $a(F) \leq n+1$.

Moreover, if a convex body F in R^n has at most n non-regular boundary points, then $a(F) \leq n+1$.

Solutions to Exercises

17.1. Let pqr be a triangle and h its interior point (Figure 17.13). We will show that the three light beams defined by vectors $\overrightarrow{ph}, \overrightarrow{qh}$, and $\overrightarrow{rh}$ illuminate the boundary of each convex figure F whose boundary contains only regular points.

Indeed, let x be a boundary point of F.Consider the triangle $p'q'r'$, the image of pqr under translation through vector $\overrightarrow{hx}$ (Figure 17.14). Let L be the support line of F through x (note that it is unique because the boundary point x is regular). Then at least one of the points p', q', and r', say q', is situated on the opposite side of L from F. Now it is clear that the beam of direction $\overrightarrow{q'x} = \overrightarrow{qh}$ illuminates x, that is, the

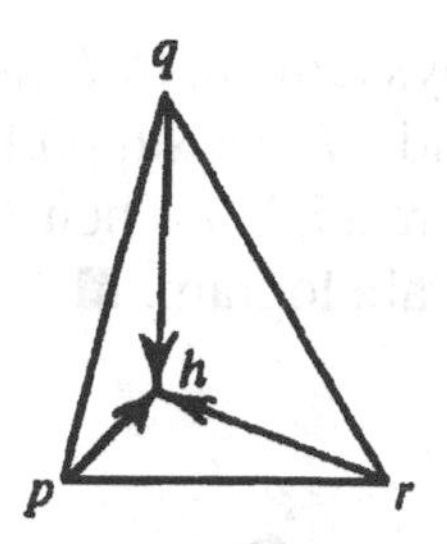

Fig. 17.13

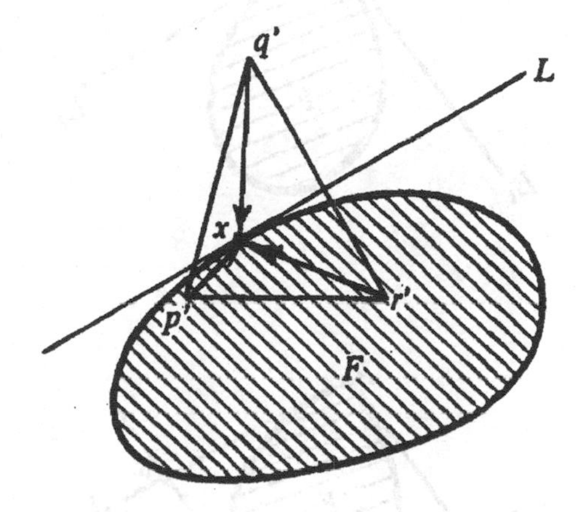

Fig. 17.14

line L' through q' and x intersects the interior of F. Indeed, otherwise L' and L (Figure 17.15) would be distinct support lines of F at x, contradicting the assumption that x is a regular boundary point. ■

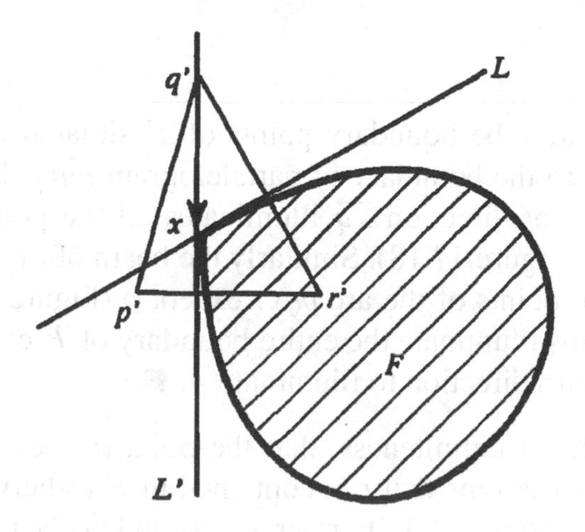

Fig. 17.15

17.2. Let L_1 and L_2 be rays tangent to F at boundary point p (Figure 17.16). Further, let M_1 and M_2 be support lines of F parallel to L_1 and L_2, respectively (Figure 17.17). Then L_1 and L_2 and M_1 and M_2 determine the required parallelogram. ■

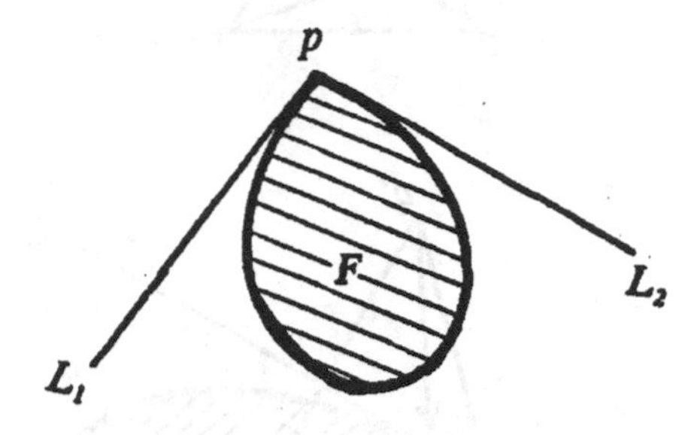

Fig. 17.16

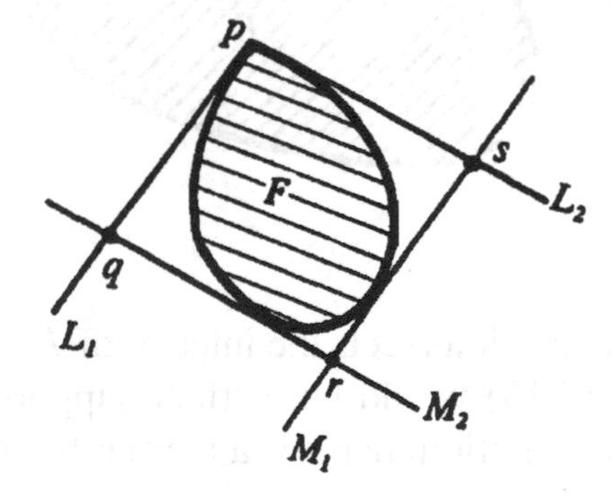

Fig. 17.17

17.3. Let x and y be boundary points of F situated on arc qs that do not belong to the boundary of parallelogram $pqrs$ (Figure 17.18). Then the beam of direction $\overrightarrow{xq}$ illuminates all the points of the arc $psyx$ except p (Figure 17.18). Similarly the beam of direction $\overrightarrow{ys}$ illuminates all the points of the arc $pqxy$ except p (Figure 17.19). Thus, these two beams illuminate the entire boundary of F except the point p. It takes a third direction to illuminate p. ■

17.4. Assume, for definiteness, that the point q does not belong to F. We may also assume that r is contained in F (otherwise the result follows from Exercise 17.3). Further, let m_1 and m_2 be rays emanating

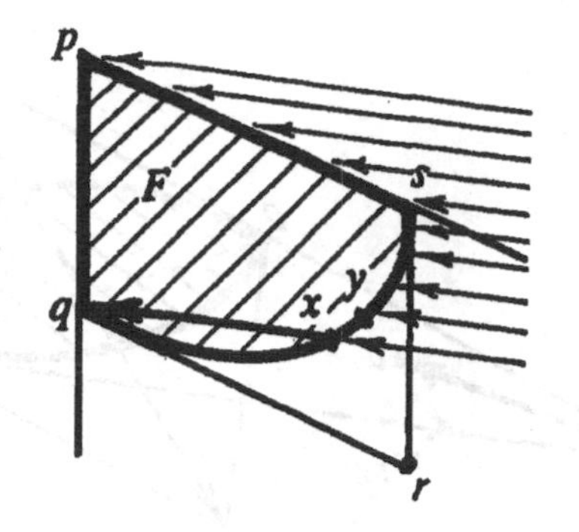

Fig. 17.18

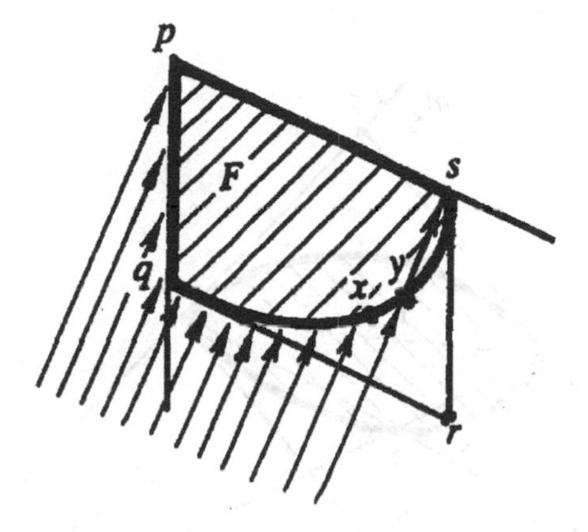

Fig. 17.19

from r and tangent to F at r. First we consider the case when at least one of m_1, m_2, say m_1, is situated in interior of angle qrs (Figure 17.20).

Let us denote by u the point of intersection of the line ps and the ray m_1. Further, we choose an interior point d of the angle sur. Then the beam of direction $\overrightarrow{ud}$ illuminates all the points of the arc pr (situated in the triangle pqr) including p but excluding r. The beam of direction $\overrightarrow{sq}$ illuminates all the points of the arc pr (situated the triangle spr) except endpoints p and r. Finally, if e is an interior point of the angle between m_1 and m_2, then the beam of direction $\overrightarrow{re}$ illuminates the point r. Thus, the entire boundary of F is illuminated by three light beams of directions $\overrightarrow{ud}$, $\overrightarrow{sq}$, and $\overrightarrow{re}$.

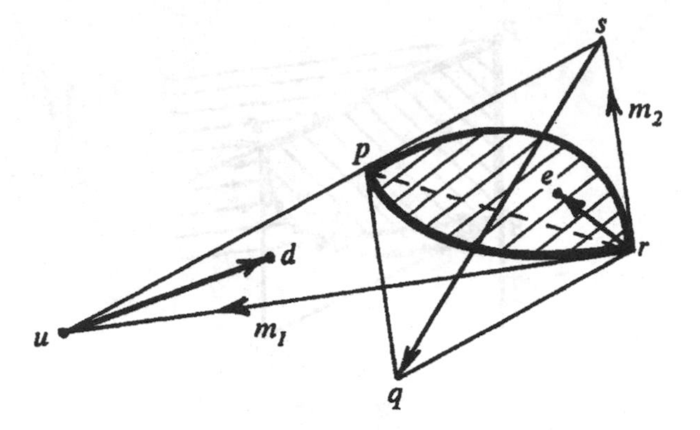

Fig. 17.20

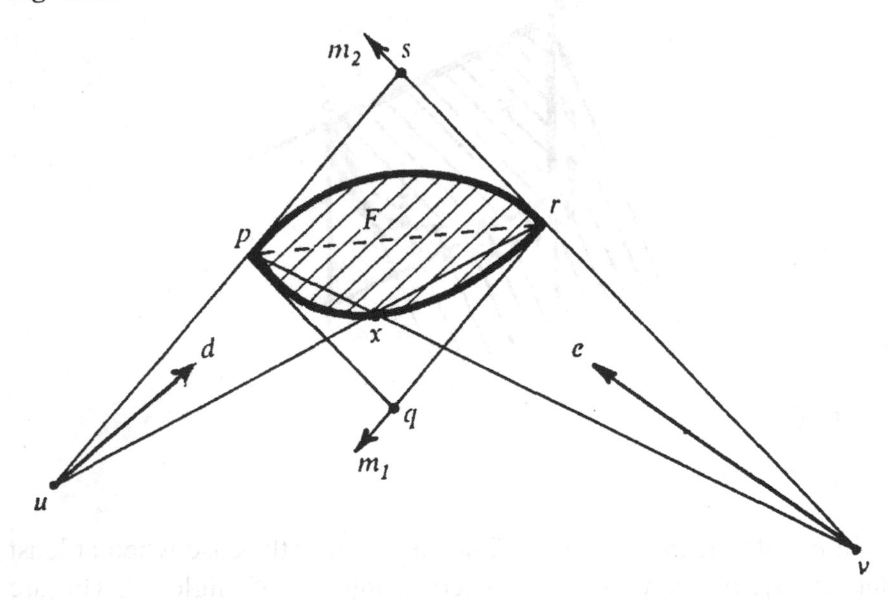

Fig. 17.21

It remains to consider the case when rays m_1 and m_2 pass through the points q and s, respectively. Let x be a boundary point of F situated in the arc pr (in the triangle pqr) that does not belong to the boundary of the parallelogram $pqrs$ (Figure 17.21). Then x does not belong to the diagonal $[p,r]$ (since the ray emanating from p and passing through q is tangent to F). Let us denote the point of intersection of the lines sp and rx by u, and the point of intersection of the lines sr and px by v. Further, we choose an interior point d of the angle sur and an interior point e of the angle svp.

Then the beam of direction $\overrightarrow{ud}$ illuminates all the points of the arc px (including endpoints) and the beam of direction $\overrightarrow{ve}$ illuminates all the points of the arc rx (including endpoints). Finally, the beam of direction $\overrightarrow{sq}$ illuminates all the points of the arc pr (in the triangle pry) except the endpoints. So, the beams of directions $\overrightarrow{ud}$, $\overrightarrow{ve}$, and $\overrightarrow{sq}$, illuminate the whole boundary of F in this case. ■

17.5. Exercises 17.1, 17.2, 17.3, and 17.4 give a proof of Theorem 17.1. Indeed, if F is smooth, then $c(F) = 3$ (Exercise 17.1). Let us assume now that the boundary of F contains at least one angular point p. We consider the parallelogram $pqrs$, as in Exercise 17.2 (see Figure 17.5). If r does not belong to F, then $c(F) = 3$ (Exercise 17.3). If r is contained in F but at least one of the points q or s does not belong to F, then $c(F) = 3$ as well (Exercise 17.4). Finally, when all the points p, q, r, and s are contained in F, F coincides with the parallelogram $pqrs$. ■

17.6. The reasoning is similar to the solution of Exercise 17.1. We consider a tetrahedron $pqrs$ and its interior point h. Then the beams of directions $\overrightarrow{ph}, \overrightarrow{qh}, \overrightarrow{rh}$, and $\overrightarrow{sh}$ illuminate the boundary of any smooth convex body F. Indeed, let $p'q'r'\,s'$ be the image of $pqrs$ under translation through the vector $\overrightarrow{hx}$, where x is a boundary point of F. Let L be the support plane of F through x. Then at least one of the points p', q', r', and s', say q', is situated on the other side of the plane L from F. Now it is clear that the beam of direction $\overrightarrow{q'x} = \overrightarrow{qh}$ illuminates the point x, that is, the line l' through q' and x intersects the interior of F. Indeed, otherwise there exists a support plane L' containing l'. L' does not coincide with L, since L does not contain l'. Thus, there are two support planes, L and L', through x, contradicting the assumption that the boundary point x is regular. ■

17.7. Let us consider three beams that illuminate the entire boundary of F (Figure 17.22). Then the tangent points b and c are distinct, and the tangent points d and e are distinct, too (see notation in Figure 17.22), and the tangent points d and a are distinct. We choose points p, q, and r between these pairs of distinct points, respectively, and denote an interior point of F by g. Then F is the union of three figures P_1, P_2, and P_3, each of which is the convex hull of g and one of the

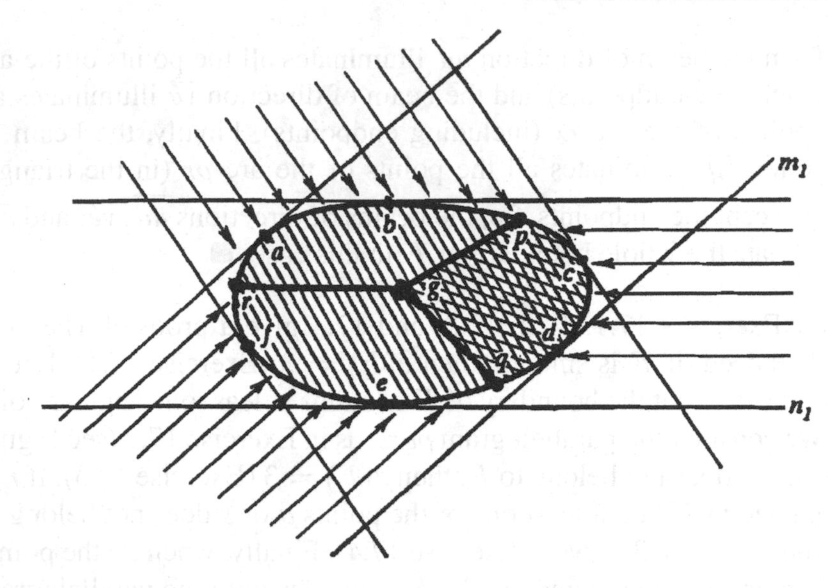

Fig. 17.22

arcs pq, qr, and rp (the figure P_1 is shaded twice in Figure 17.22). Notice that P_1 is situated *strictly between* support lines m_1 and k_1. So if a point o_1 is situated far enough to the right (between support lines at b and e parallel to n_1) and the ratio k_1 is close enough to 1, but less than 1, then the homothety h_1 with center o_1 and ratio k_1 maps F onto figure $h_1(F)$ that contains P_1.

By similarly choosing homotheties h_2 and h_3, we obtain the required smaller copies $h_1(F)$, $h_2(F)$, $h_3(F)$ of F.

This reasoning can be generalized to the n-dimensional case. ■

17.8. Let q be a point that belong to F (Figure 17.23). Then there exist two lines through q tangent to the parabola P. Denote the corresponding support points by m, n. These points bound the arc mn of finite length. If now F_1 is a "smaller copy" of F, that is, the image of F under a homothety with center q and positive ratio $k < 1$, then the intersection of F_1 with the parabola P is an arc contained in mn. Thus, each "smaller copy" covers only a finite length arc of P, and, consequently, it takes infinitely many "smaller copies" of F to cover F. This means that $b(F)$ is infinite. ■

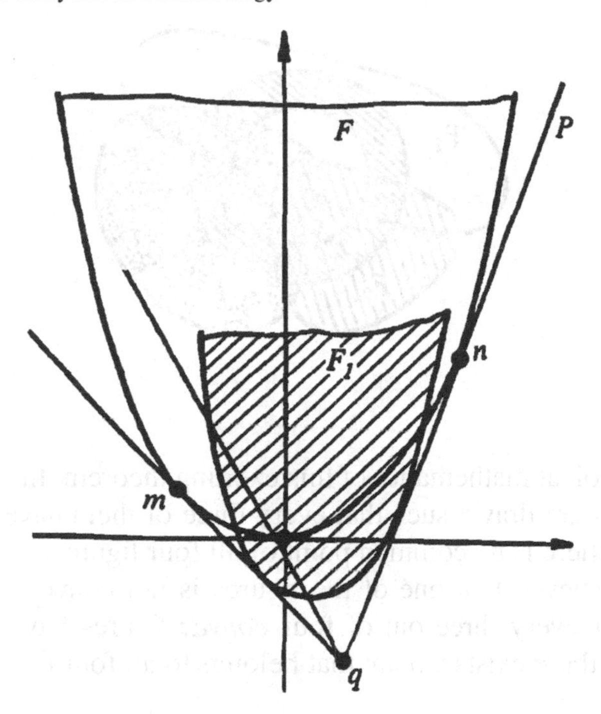

Fig. 17.23

18 Theorems of Helly and Szökefalvi-Nagy

The first of the theorems to be discussed in this section was discovered by the Austrian mathematician Eduard Helly (Vienna, 1884 – Chicago, 1943), in 1913. But it was not published at that time. During World War I, Helly was a soldier in the Austrian Army, and he was taken a Russian prisoner in 1914. He explained his theorem to another Austrian mathematician who was in Russian captivity as well.

The theorem lived in mathematical folklore from the time of this mysterious meeting of two mathematical prisoners until 1921, when the first proof of the Helly Theorem was published by the Austrian mathematician Johann Karl August Radon (with a reference to Helly). In 1923 Helly published his own proof (different from Radon's!). Dozens of proofs of the Helly Theorem are known today. Numerous papers have been written in which the theorem is applied.

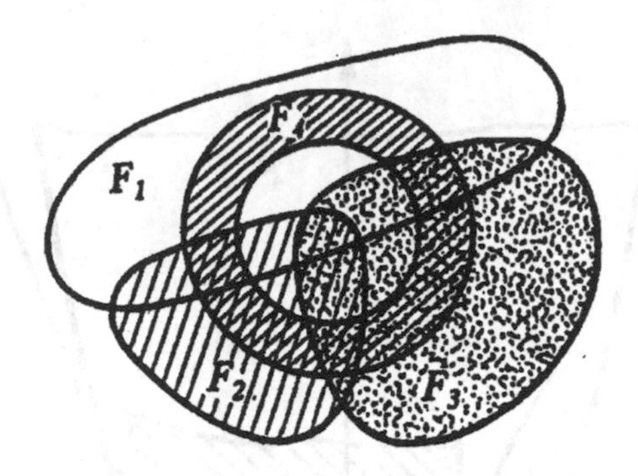

Fig. 18.1

Let us look at mathematics of this exciting theorem. In Figure 18.1, four figures are drawn such that every three of them have a common point. But there is no common point of all four figures.

Note, however, that one of the figures is not convex. And this is essential: if every three out of four *convex* figures have a common point, then there exists a point that belongs to all four figures.

Exercise 18.1. Four convex figures are given in the plane. Prove that if every three of them have a nonempty intersection, then the intersection of all four figures is nonempty as well.

Exercise 18.2. (The Helly Theorem for the Plane) A finite family $F_1, \ldots, F_s$ of convex figures is given in the plane. Prove that if every three of them have a nonempty intersection, then $F_1 \cap \ldots \cap F_s$ is nonempty as well.

Note that in the statement of Exercise 18.2 we assume that every *three* of the given convex figures have a nonempty intersection. In the following theorem, three is equal to $n + 1$, where $n = 2$ is the *dimension* of the plane. This suggests the statement of the Helly Theorem in space (that has dimension $n = 3$): *if every four of the given convex figures $F_1, \ldots, F_s$ in 3-space have a common point, then the intersection of all of the figures is nonempty.*

Here is the statement of the theorem for n-dimensional space R^n:

The Helly Theorem 18.1. Let $F_1, \ldots, F_s$ be convex figures in R^n. If every set of $n + 1$ of these figures has a common point, then the $F_1 \cap \ldots \cap F_s$ is nonempty.

Certainly, we assume in this statement that s is a large number, at any rate, that $s \geq n + 1$. The statements for the plane (Exercise 18.2) and for 3-dimensional space are particular cases of this general theorem.

What is the "profit" of the Helly Theorem? From numerous examples of its application we will list several in the following exercises.

Exercise 18.3. Given s points in the plane ($s > 0$) such that every three of them are contained in a disk of radius d. Prove that all the given points are contained in a disk of radius d.

Exercise 18.4. (Theorem of H. Jung) Let M be a convex polygon in the plane such that the distance between every two of its vertices does not exceed d. Prove that M is contained in a disk of diameter $d\frac{2}{\sqrt{3}}$.

In the above statements we considered only *finite* families of convex figures. It is interesting to ask whether the Helly Theorem holds for an *infinite* family of convex figures. The answer is "yes," if all the figures are closed and one of them is *compact* (that is, closed and bounded).

Exercise 18.5. Let $F_1, \ldots, F_s \ldots$ be a sequence of convex figures in the plane, each of which is closed and at least one of which is compact. Prove that if every three of the figures $F_1, \ldots, F_s \ldots$ have a nonempty intersection, then the intersection $F_1 \cap \ldots \cap F_s \ldots$ of all of the figures is nonempty as well.

Exercise 18.6. Is the condition "each of the figures is closed" in Exercise 18.5 essential?

Exercise 18.7. Is the condition "one of the figures is compact" in Exercise 18.5 essential?

In Exercise 18.5, we had an infinite sequence of convex figures. But the assertion of this exercise holds for every family of convex figures, not only those enumerated by positive integers:

The Helly Theorem 18.2 for an Infinite Family of Convex Figures. Given an infinite family of closed convex figures in the plane, one of which is compact, and such that every three of them have a common point. Then the intersection of all figures of the family is nonempty.

The statement for R^3 is similar.

There is a beautiful conjecture associated with the above theorem. The great Paul Erdős suggested to Alexander Soifer to include it here for your (and our) enjoyment. Thank you, Paul!

Erdős's Conjecture 18.3. Given an infinite family of closed convex figures in the plane, one of which is compact. If among any four figures there are three figures with a point in common, then there is a finite set S (consisting of N points) such that every given figure contains at least one point from S.

Moreover, the positive integer N is an absolute constant, i.e., it is one and the same for all families of figures that satisfy the above conditions.

To make it more exciting, we offered $25 for the first proof of Erdős's Conjecture 18.3 or a counterexample disproving it.

18 years later[1], in late September 2008, while reading the manuscript of this new edition, Branko Grünbaum resolved this conjecture in the negative: and won the $25 prize. Grünbaum showed that Erdős's Conjecture 18.3 does not hold even for the line of real numbers R.

Grünbaum's Counterexample 18.4. (e-mail to A. Soifer, September 26, 2008) Define the sets as follows:

$$F_0 = \{0\};$$
$$F_n = \{x \in R : x \geq n\} \text{ for every positive integer } n.$$

All conditions of Erdős's Conjecture 18.3 are satisfied, while for any finite set S of real numbers, there is an integer n that is greater than any number from S. By definition, F_n does not contain any element from S. ■

On September 29, 2008, I asked Grünbaum whether he could "save" Erdős's Conjecture 18.3, and the following day he sent me his solution:

> Yes, I conjecture that Erdős's problem may be resuscitated by requiring two (instead of just one) of the sets to be compact. But I do not see any easy proof.

[1] This train of thought was added for this new Springer 2010 edition.

Grünbaum's Conjecture 18.5. (e-mail to A. Soifer, September 30, 2008) Given an infinite family of closed convex figures in the plane, *two* of which are compact. If among any four figures there are three figures with a point in common, then there is a finite set S (consisting of N points) such that every given figure contains at least one point from S.

Due to the Helly Theorem for an infinite family of convex figures, the statements of Exercises 18.3 and 18.4 can be generalized.

Exercise 18.3. Let M be a figure in the plane such that any three of its points are contained in a disk of radius d. Then M is contained in a disk of radius d.

Exercise 18.4. Each figure of diameter d in the plane is contained in a disk of diameter $\frac{2}{\sqrt{3}}d$.

Exercise 18.8. Given a bounded figure M of area S is in the plane. Prove that there exists a point q in the plane such that each line through q divides M into two parts, each of which has area at least $\frac{1}{3}S$ (Figure 18.2).

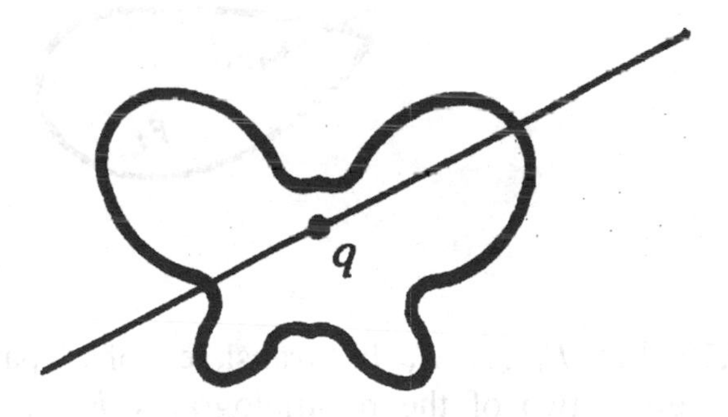

Fig. 18.2

Exercise 18.9. Given a bounded convex figure M. Prove that there exists an interior point q of M such that every chord $[ab]$ of M through q is divided by q into two parts, each of which is of length at least $\frac{1}{3}|ab|$ (Figure 18.3).

In the following exercises we consider *translates* of a convex figure F, that is, figures obtained from F by translations (Figure 18.4).

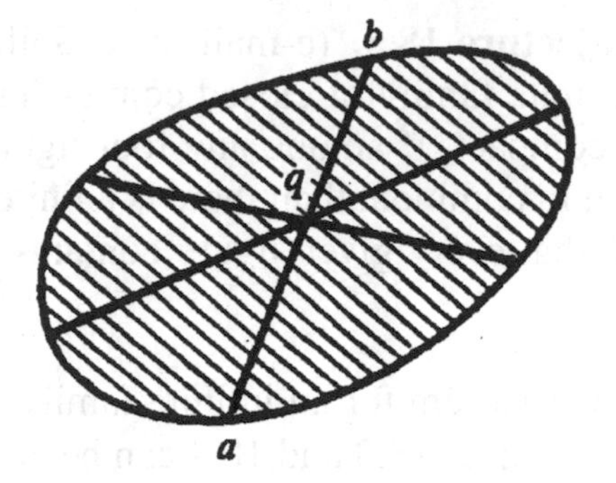

Fig. 18.3

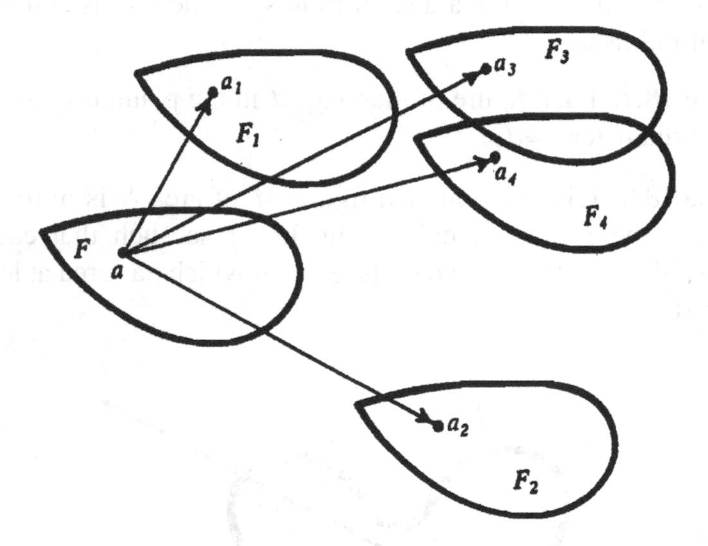

Fig. 18.4

Exercise 18.10. Let $F_1, \ldots, F_s$ be translates of a parallelogram. Prove that if every two of the parallelograms $F_1, \ldots, F_s$ have a point in common, then the intersection of all the parallelograms is nonempty.

What is the difference between the Helly Theorem and the previous exercise? In the Helly Theorem we assume that every *three* of convex figures $F_1, \ldots, F_s$ have a common point and conclude that the intersection of all of the figures is nonempty. In Exercise 18.10 we have a special case when $F_1, \ldots, F_s$ are translates of a parallelogram. In this case, it suffices to assume that every *two* of the figures $F_1, \ldots, F_s$

have a common point for us to conclude that the intersection of all figures is nonempty.

Is there a convex figure F (not a parallelogram) with the same property: for any of its translates $F_1, \ldots, F_s$, if any two of them have a point in common then $F_1 \cap \ldots \cap F_s$ is nonempty? The answer is "no." Confirming this fact is the next task for the reader:

Exercise 18.11. Prove that if F is a convex figure in the plane (not a parallelogram), then there exist three translates F_1, F_2, F_3, of F that pairwise have points in common, but their intersection $F_1 \cap F_2 \cap F_s$ is empty.

The statement of the previous exercise is a particular case of a theorem that was established by the well-known Hungarian mathematician Béla Szökefalvi-Nagy (1913–1998) in 1954. Let us formulate this result.

The Szökefalvi-Nagy Theorem 18.6. Let F be a convex figure in R^n with the following property: for any of its translates $F_1, \ldots, F_s$, if any two of them have a point in common then the intersection $F_1 \cap \ldots \cap F_s$ is nonempty. Then F is a parallelotope (i.e., the n-dimensional analog of a parallelepiped in R^3).

This theorem raises an interesting problem:

The Szökefalvi-Nagy Problem 18.7. Let r and n be positive integers and $r \leq n$. Describe all convex bodies F in R^n with the following property: for any translates $F_1, \ldots, F_s$ of F, if any $r+1$ of them have a point in common, then $F_1 \cap \ldots \cap F_s$ is nonempty.

We will denote this property by $Sz(r)$.

In the case $r = n$, the solution is given by the Helly Theorem: $Sz(n)$ holds for every convex body in R^n. In the case $r = 1$, the solution is given by the Szökefalvi-Nagy Theorem: $Sz(1)$ holds if and only if F is a parallelotope in R^n. It would be great to solve this problem for the intermediate cases: $1 < r < n$.

The Szökefalvi-Nagy Problem is not completely unsolved. Some partial results have been established. A solution was obtained for centrally symmetric convex bodies by V. G. Boltyanski in 1976 [B3] in the case $n = 2$, and by the Hungarian mathematician J. Kincses in 1987 [Ki] in the cases $n = 3$ and $n = 4$. A solution was recently obtained by V. G. Boltyanski [B6] without the assumption of central symmetry. This means that in R^3 we have a complete solution (that is, for $r = 1, 2$ or 3). But for $n > 3$, a complete solution is still unknown.

In conclusion, we would like to describe some interesting facts about R^3 obtained in [B3] and [B6].

A centrally symmetric body F in R^3 possesses the property $Sz(2)$ if and only if it is *a cylinder,* that is, the vector sum of a two-dimensional convex figure and a segment (Figure 18.5). In other words, let F be a centrally symmetric convex body in R^3 with the following property: for any translates $F_1, \ldots, F_s$ of F, if any three of them have a point in common, then $F_1 \cap \ldots \cap F_s$ is nonempty. Then F is a cylinder with a centrally symmetric base.

Another interesting result is related to pyramids with quadrangular base (Figure 18.6). If F_0 is a pyramid with a parallelogram as its base, then F_0 possesses the property $Sz(2)$. In other words, if any three of the translates $F_1, \ldots, F_s$ of to have a common point, then $F_1 \cap \ldots \cap F_s$ is nonempty.

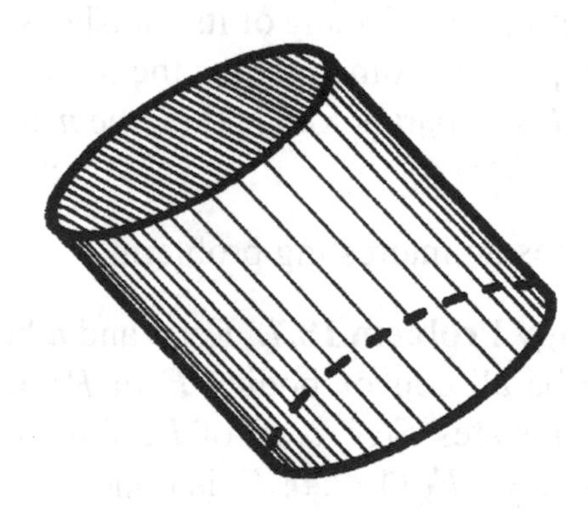

Fig. 18.5

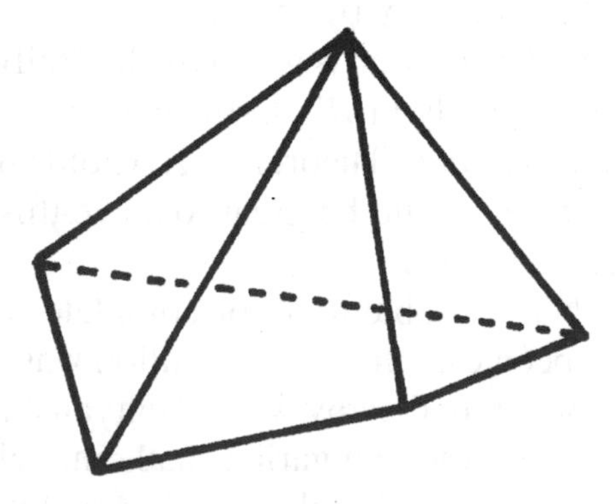

Fig. 18.6

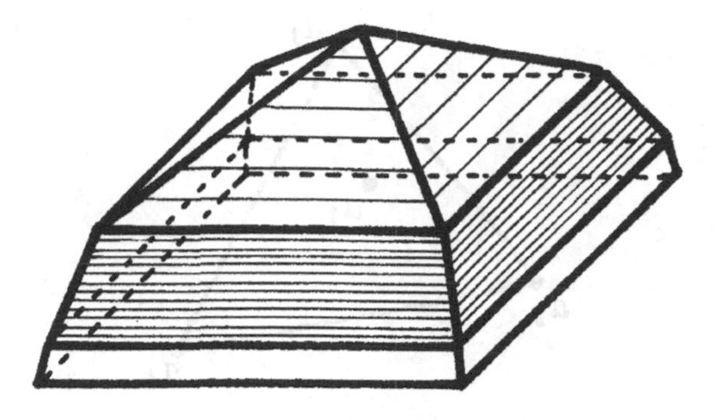

Fig. 18.7

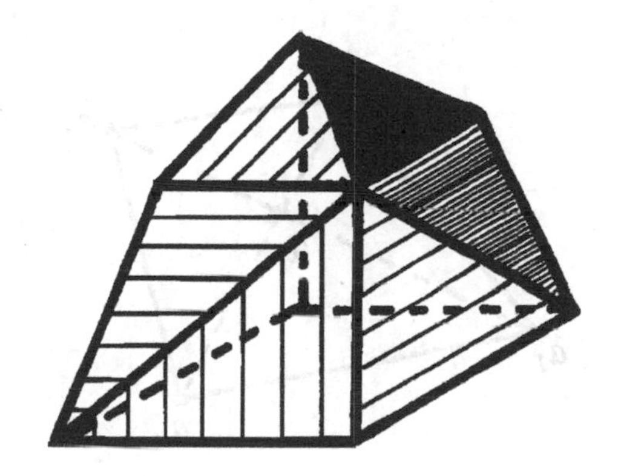

Fig. 18.8

But if the base of a pyramid F is a quadrangle distinct from a parallelogram, this assertion is not true! This means that there are four translates F_1, F_2, F_3, and F_4 of F such that any three of them have a common point, but $F_1 \cap F_2 \cap F_3 \cap F_4$ is empty (can you prove it?).

In Figures 18.7 and 18.8, some other convex bodies with the property $Sz(2)$ are shown.

Solutions to Exercises

18.1. Let a_j be a point that belongs to all convex figures F_1, F_2, F_3, F_4, except possibly F_j. We consider two cases:

i) One of the points a_1, a_2, a_3, or a_4 is contained in the triangle with vertices at the other three points, say a_1 is contained in $a_2a_3a_4$ (Figure 18.9).

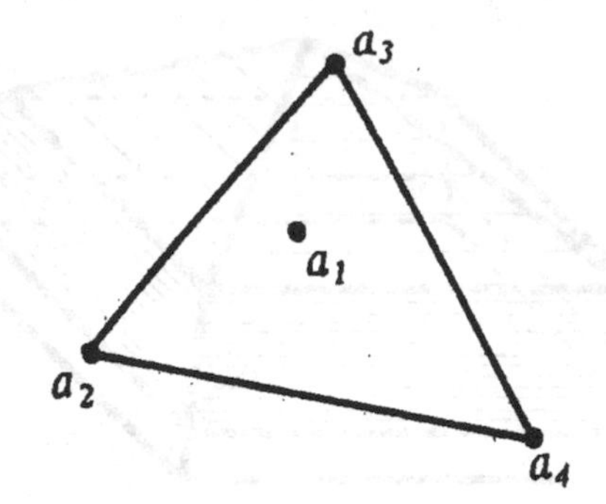

Fig. 18.9

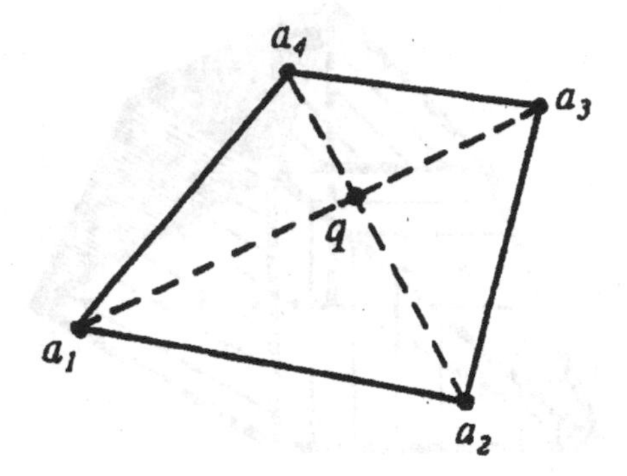

Fig. 18.10

Each of the points a_2, a_3, a_4 belongs to the figure F_1. Consequently, $a_2a_3a_4$ is contained in F_1. Hence, a_1 is contained in F_1, too. Moreover, a_1 belongs to each of F_2, F_3, F_4. Thus a_1 is a common point of F_1, F_2, F_3, F_4.

ii) None of the points a_1, a_2, a_3, a_4 is contained in the triangle with vertices at other three points, that is, a_1, a_2, a_3, and a_4 are the vertices of a convex quadrangle (Figure 18.10).

Without loss of generality we can assume that a_1 and a_3 are the opposite vertices of the quadrangle. a_1 belongs to each of the figures F_2 and F_4. a_3 belongs to F_2 and F_4, too. Consequently, the segment $[a_1, a_3]$ is contained in F_2 and F_4. Similarly, the segment $[a_2, a_4]$ is contained in F_1 and F_3. Thus, the intersection point of the diagonals $[a_1, a_3]$ and $[a_2, a_4]$ belongs to each of the figures F_1, F_2, F_3, and F_4.

Cases (i) and (ii) address all possible situations when the points a_1, a_2, a_3, a_4 are not all on a line. If they are all on a line, then the proof is not difficult (try it on your own!). ■

18.2. Let us proceed by induction on s. If $s = 4$, then the assertion is true (Exercise 18.1). Assume that for $s \geq 4$ the assertion is true. For $s + 1$ convex figures $F_1, \ldots, F_s, F_{s+1}$, such that any three of them have a nonempty intersection, we denote $F_j \cap F_{s+1}$ by G_j (where $j = 1, \ldots, s$). Then any three of the convex figures $G_1, \ldots, G_s$ have a common point (indeed, $G_i \cap G_j \cap G_k$ is equal to $F_i \cap F_j \cap F_k \cap F_{s+1}$, and this intersection is nonempty, by Exercise 18.1). Hence, by the inductive assumption, $G_1 \cap \ldots \cap G_s$ is nonempty. But this intersection coincides with $F_1 \cap \ldots \cap F_s \cap F_{s+1}$. ■

18.3. We have to prove that there exists a point q (the center of the desired disk) such that the distance from each given point to q does not exceed d. In other words, we have to find a point q that belongs to each disk of radius d with center at one of the given points.

According to the Helly Theorem, it is sufficient to prove that any *three* of these disks have a common point. But we know that every three given points a, b, c are contained in a disk of radius d (Figure 18.11). The center x of this disk is the required common point of the disks of radius d with centers at a, b, and c. ■

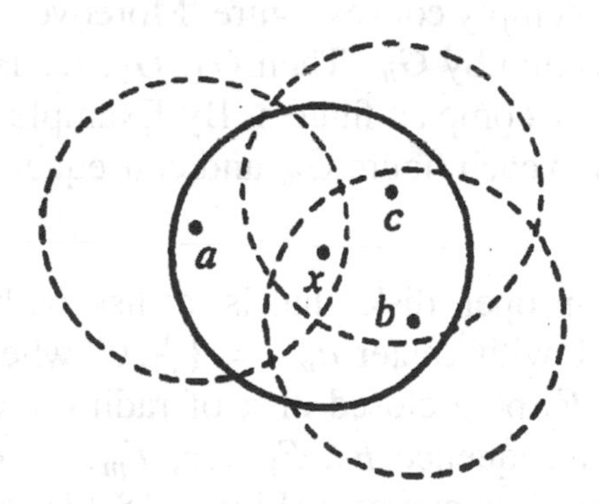

Fig. 18.11

18.4. By Exercise 18.3, it suffices to prove that any three vertices of the polygon M are contained in a disk of diameter $d\frac{2}{\sqrt{3}}$. Let $a, b,$

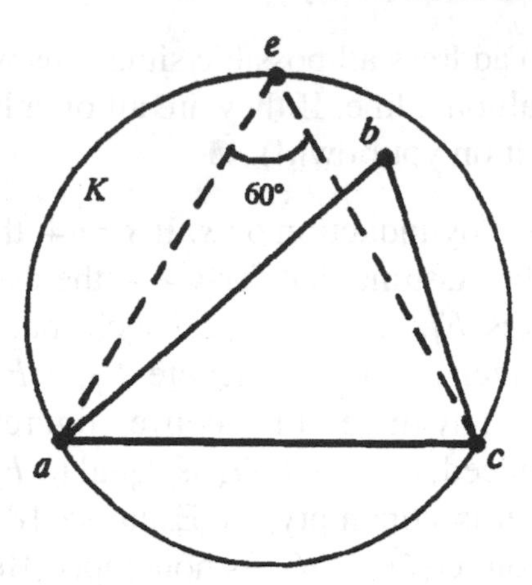

Fig. 18.12

and c be three vertices of M, and $[a, c]$ the largest side of triangle abc. Then angle abc is at least 60°. So, point b belongs to the disk K circumscribed about the equilateral triangle with side $[a, c]$ (Figure 18.12). But the diameter of this disk is equal to $|ac| \frac{2}{\sqrt{3}}$, that is, it does not exceed $d \frac{2}{\sqrt{3}}$. ■

18.5. Without loss of generality, we can assume that figure F_1 is compact. According to the Helly Theorem, the intersection of the figures $F_1, \ldots, F_m$ is a nonempty convex figure. Moreover, it is compact. We denote this intersection by G_m. Then $G_1, G_2, \ldots$ is a decreasing sequence of nonempty compact figures. By Example 8.4 there exists a point that belongs to each figure G_m and consequently, to each figure F_m. ■

18.6. Let F_m be an open disk (that is, a disk without its boundary points) of radius 1 with center $a_m = \left(\frac{1}{m}, 0\right)$ where $m = 1, 2, \ldots$ Further more, let F_0 be a closed disk of radius 1 with center $a_0 = (2, 0)$. We obtain a sequence $F_0, F_1, \ldots, F_m, \ldots$ of convex figures in the plane where F_0 is compact (Figure 18.13). Any finite number of figures $F_0, F_1, \ldots, F_m, \ldots$ has a nonempty intersection, but the intersection of *all* these figures is empty.

Note that the intersection of the corresponding *closed* disks consists of only the point (1,0). ■

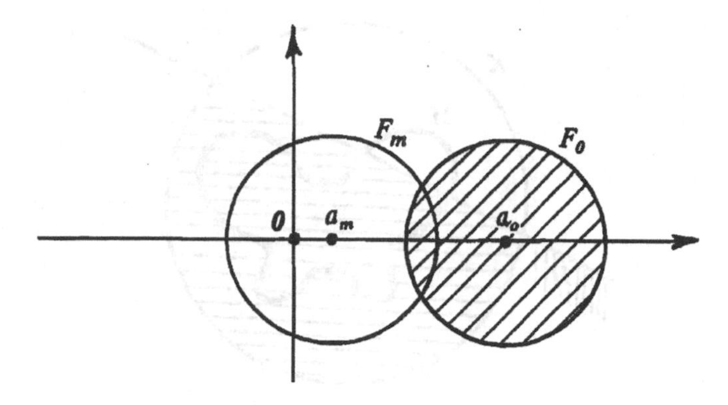

Fig. 18.13

18.7. Let us consider the family of all "right" closed half-planes defined by vertical lines (Figure 18.14). None of these figures is compact and the intersection of all figures of this infinite family is empty (since there is a half-plane moved as far away to the right as we like in the family). ■

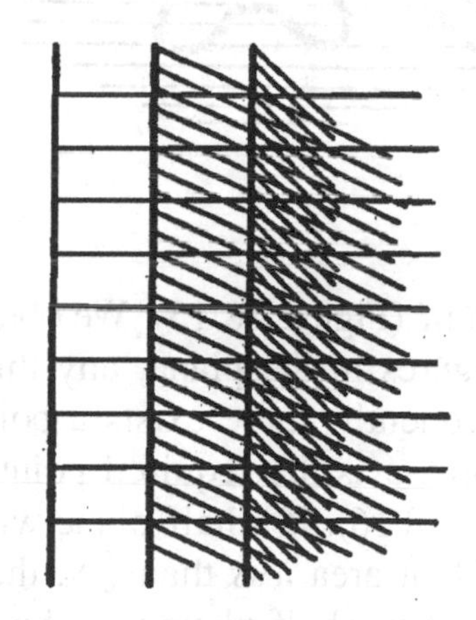

Fig. 18.14

18.8. Let us select all closed half-planes, that contain a part of M of area exceeding $\frac{2}{3}S$, so that the part of M situated outside of such a half-plane is of area *less* than $\frac{1}{3}S$. This means that any three of the selected half-planes have a nonempty intersection (can you see why?). Now let us take a disk K containing M and replace each half-plane

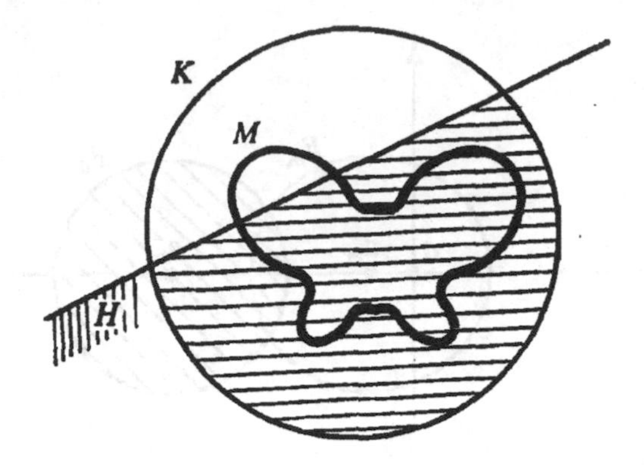

Fig. 18.15

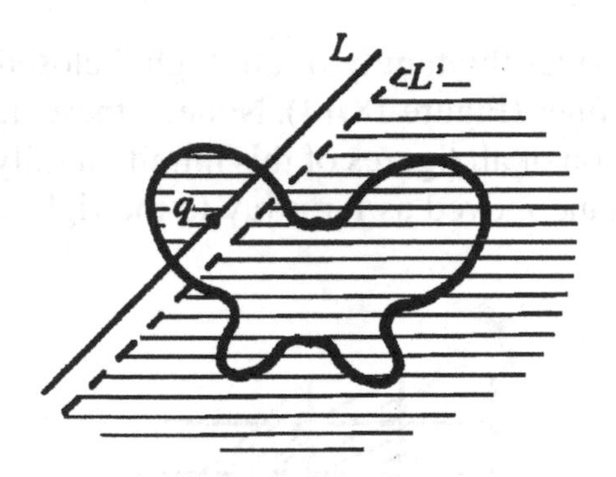

Fig. 18.16

H of our family by $H \cap K$ (Figure 18.15). We obtain an infinite family of compact convex figures in the plane, any three of which have a common point. Consequently, there exists a point q that belongs to all figures of the family. q is the required point. Indeed, let L be a line through q (Figure 18.16). If a half-plane with boundary L were to contain a part of M of area *less* than $\frac{1}{3}S$, then there would exist a parallel line L' such that a half-plane with boundary L' contains a part of M of area greater than $\frac{2}{3}S$ and does contain q, contradicting the construction of q. This completes the solution.

Note that the number $\frac{1}{3}$ in the statement of Exercise 18.3 cannot be replaced by a greater number. For example, if M consists of three separate congruent disks (Figure 18.17), then the number $\frac{1}{3}$ cannot be improved. On the other hand, if M is convex, then $\frac{1}{3}$ can be replaced by $\frac{4}{9}$. ■

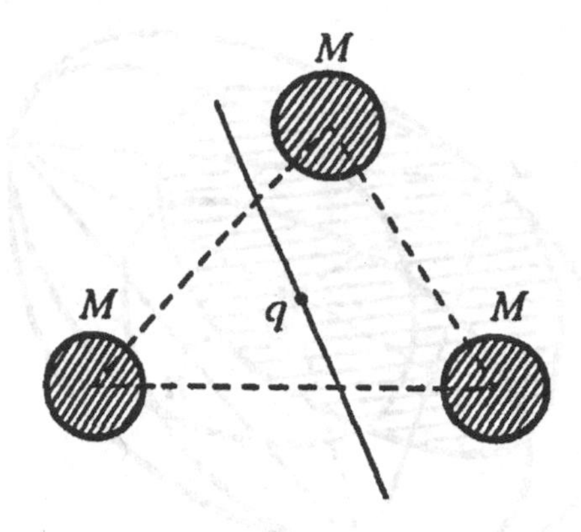

Fig. 18.17

18.9. For every point a of M we define M_a as the figure homothetic to M with center a and ratio $\frac{2}{3}$ (Figure 18.18).

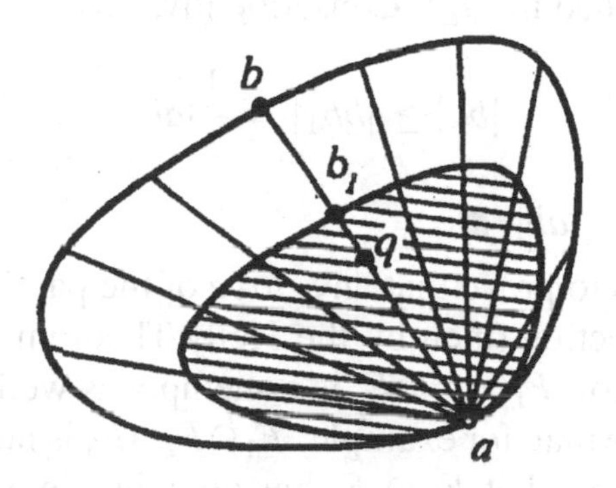

Fig. 18.18

We consider the family of all such figures M_a. If M_a, M_b, M_c, and are three figures of this family (Figure 18.19), then triangle abc is contained in M.

The median $[a, m]$ is contained in M and, consequently, the centroid p of abc (i.e., the point of intersection of medians) is contained in M_a. Similarly, p belongs to M_b and M_c. Thus, any three figures of the considered family have a common point. Moreover, each figure M_a is convex and compact. Hence, there exists a point q that belongs to all figures M_a of our family.

Now let $[a, b]$ be a chord through q (Figure 18.18), and let b_1 be the image of b under the homothety with center a and ratio $\frac{2}{3}$. Then

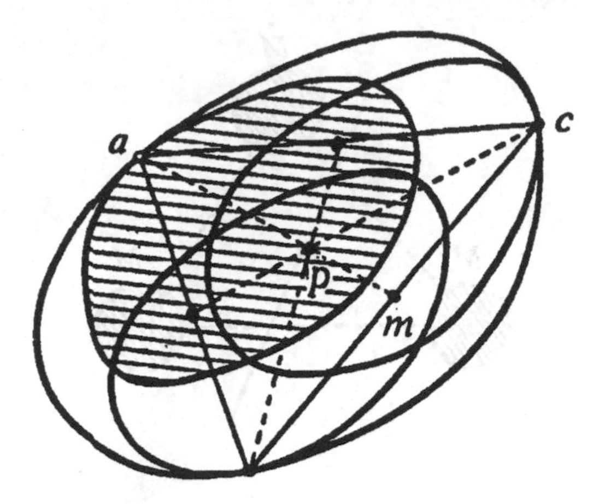

Fig. 18.19

$[a, b_1]$ is a chord of M_a, and the point q belongs to the segment $[a, b_1]$ (since q is contained in M_a). Consequently,

$$|bq| \geq |bb_1| = \frac{1}{3}|ab|.$$

Similarly, $|aq| \geq \frac{1}{3}|ab|$. ■

18.10. It suffices to prove that any *three* of the parallelograms have a nonempty intersection (then by the Helly Theorem we can conclude that the intersection $F_1 \cap \ldots \cap F_s$ is nonempty as well).

So, let us prove that, for example, $F_1 \cap F_2 \cap F_3$ is nonempty. Assume the opposite, that is, that $F_1 \cap F_2$ has no points in with the parallelogram F_3. Then there exists a line L parallel to the parallelograms that separates $F_1 \cap F_2$ and F_3 (Figure 18.20).

Let us replace each parallelogram F_i by the corresponding strip S_i with sides parallel to L. Then as before, L separates $S_1 \cap S_2$ and S_3 (Figure 18.21).

But we know that F_1 and F_3 have a common point a. Further, if we take a point p of $F_1 \cap F_2$, then both a and p belong to F_1. Consequently, the segment $[a, p]$ is contained in F_1. This segment has a common point q with the line L (since a and p are situated on the different sides of L). Thus, L intersects F_1 at q. Therefore, L is contained in S_1. Similarly, L is contained in S_2. Hence L is contained in $S_1 \cap S_2$, contradicting the construction of L. This contradiction shows that $F_1 \cap F_2 \cap F_3$ is nonempty. ■

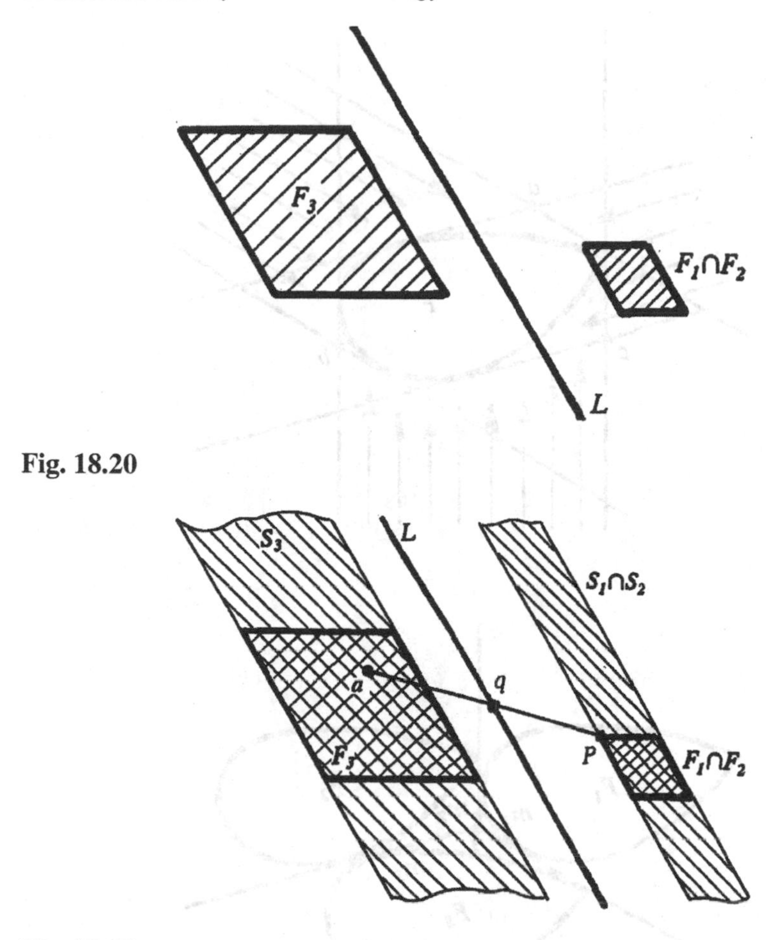

Fig. 18.20

Fig. 18.21

18.11. Since F is not a parallelogram, there are three directions that illuminate the whole boundary of F (see Theorem 17.1). Six support lines of F parallel to these illuminating directions form a hexagon circumscribed about F (Figure 18.22). The points a, b, and c in Figure 13.22 *cannot belong to* F because otherwise they would not be illuminated.

Let us now choose a point q and consider three translations through the vectors $\overrightarrow{aq}, \overrightarrow{bq}$ and $\overrightarrow{cq}$ (Figure 18.23).

We obtain three translates F_1, F_2, and F_3 of F that do not contain the point q and pairwise have a common supporting ray emanating from q. Finally, we choose three points m_1, m_2, and m_3 on these rays

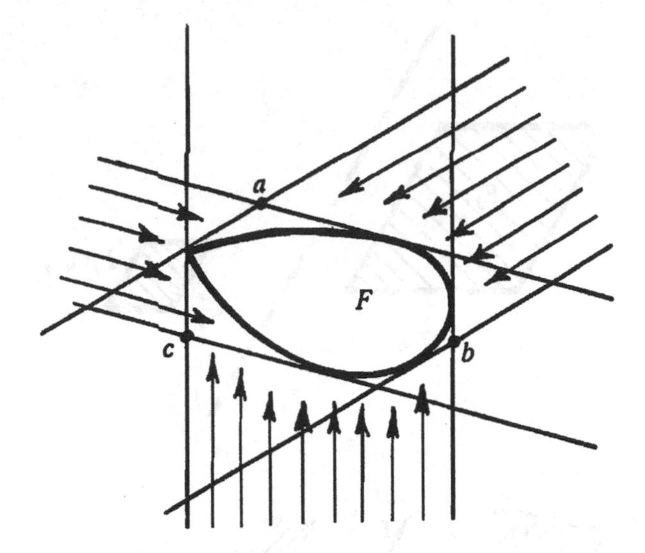

Fig. 18.22

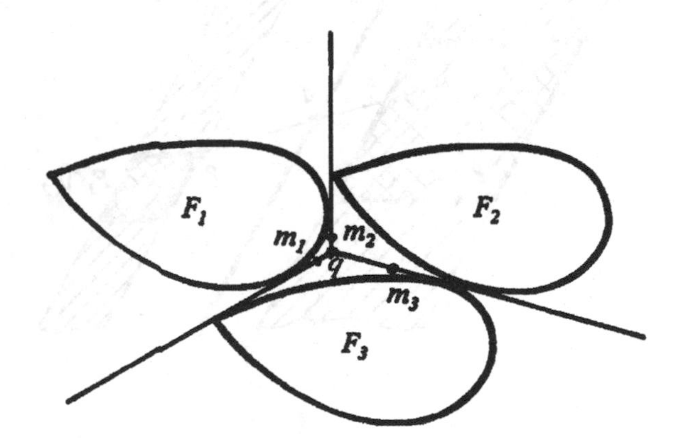

Fig. 18.23

closer to q than to F_1, F_2, F_3 and translate F_1, F_2, F_3 to new positions such that the translated figures each contain two of the points m_1, m_2, or m_3, but do not contain q (can you see why this is always possible?). These new translates of F have pairwise common points m_1, m_2, and m_3, but the intersection of all three figures is empty. ■

Part II
New Landscape, or The View 18 Years Later

Since its release 18 years ago, in 1991, *Geometric Etudes in Combinatorial Mathematics* lived its own life, quite separate from mine. It was well received by the journals that reviewed it, and by many universities that used it in classrooms and lecture halls. More importantly, the book inspired new solutions, research, results in some wonderful mathematicians, young and not so-young. You can see its influence on such papers as [Kar] and [St1].

Let us look at four new chapters containing some of these results, which continue the trains of mathematical thought you read in Part I. I have also included here the fifth new etude on my favorite open problem in all of mathematics: finding the chromatic number of the plane.

Chapter 5
Mitya Karabash and a Tiling Conjecture

In 2005, the brilliant young freshman from Columbia University Dmytro (Mitya) Karabash came to the University of Colorado at Colorado Springs, where I supervised his summer research. We looked at problems of tiling and covering, the chromatic number of the plane, and other problems posed by Paul Erdős and I. Mitya and I wrote a joint research paper on covering triangles with triangles. A year later, Mitya was first to solve Problem 5.3 posed in this book. He also proved some interesting related results.

Figures 19.1a and 19.1b show L-tetrominoes of different orientations. Figure 19.1(c) shows that 3×2 rectangle can be tiled by L-tetrominoes. In addition, the tiling in Figure 19.1(c) possesses two other properties: it has 2-way symmetry and L-tetrominoes of the same orientation.

For convenience let us call a rectangle *orientable* if it can be tiled by L-tetrominoes of the same orientation. Let us call a rectangle *symmetrically orientable* if it can be tiled by L-tetrominoes of the same orientation in such a way that tiling has 2-fold symmetry.

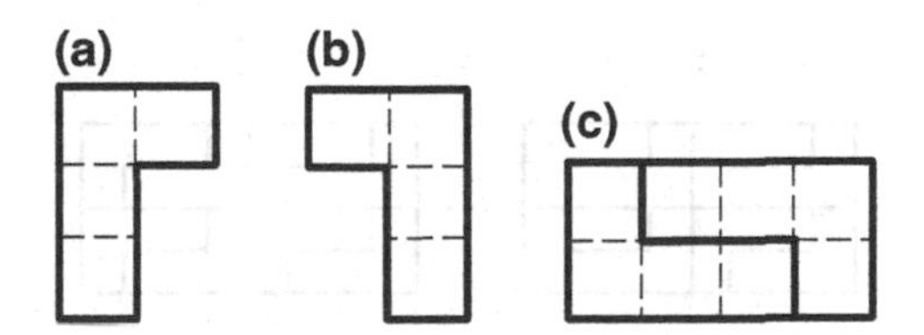

Fig. 19.1

A. Soifer, *Geometric Etudes in Combinatorial Mathematics*,
DOI 10.1007/978-0-387-75470-3_5, © Alexander Soifer, 2010

In this terminology, Problem 5.3 can be formulated as follows:

Problem 19.1. Is it true that a $m \times n$ rectangle is orientable if and only if one of the numbers m, n is divisible by 4 and the other is even?

Mitya constructed a counterexample to this conjecture.

Karabash's Counterexample 19.2. ([Kar])

Proof. Behold (Fig. 19.2):

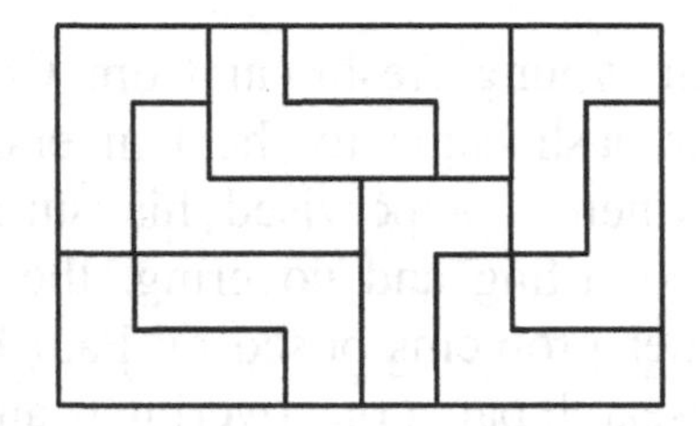

Fig. 19.2

Mitya then proceeded to classify all orientable and all symmetrically orientable rectangles (Theorems 19.7 and 19.8).

Tool 19.3. If $2|n$, $2|m$, and $8|nm$, then an $n \times m$ rectangle is orientable.

Proof. Behold (Fig. 19.3):

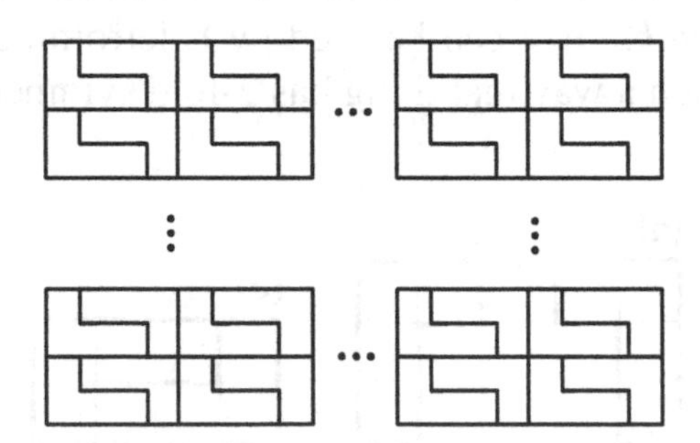

Fig. 19.3

Tool 19.4. If n is odd, $n > 3$, and $8|nm$, then the $n \times m$ rectangle is orientable.

Proof. $8|nm$ and n odd implies $8|m$. Let $m = 8k$. By Tool 19.3, the $(n-5) \times m$ rectangle is orientable. Figure 19.2 shows that the 5×8 rectangle is orientable. If $n = 5$, then divide the $n \times m$ rectangle into k 5×8 rectangles; if $n > 5$ then divide the $n \times m$ rectangle into k 5×8 rectangles and an $(n-5) \times m$ rectangle. Since the $n \times m$ rectangle can be partitioned into orientable rectangles, it must be orientable itself. ■

Tool 19.5. If the $n \times m$ rectangle is orientable, then $8|nm$.

Proof. Let k be the number of L-tetrominoes in the tiling of an $n \times m$ rectangle with L-tetrominoes of the same orientation. We will show that k must be even. Observe that $4|nm$, since the area consideration dictates $nm = 4k$. Without loss of generality, assume n is even.

Now apply cyclic row 2-coloring to the n rows of the given rectangle, i.e., alternate colors of rows between black and white. Since n is even, the number of black squares is equal to the number of white squares.

Notice that each L-tetromino covers either 3 black squares and 1 white square or 1 black square and 3 white squares. Let w be the number of L-tetrominoes that have 3 white squares and 1 black square. Then the number of white squares in the tiled rectangle is $3w + (k - w) = k + 2w$ and the number of black squares is $3(k - w) + w = 3k - 2w$. Since the number of white squares is equal to the number of black squares, we get $k + 2w = 3k - 2w$. Hence $k = 2w$ and thus k is even. ■

Tool 19.6. If the $n \times m$ rectangle is orientable, then $n \neq 1$, and $n \neq 3$.

Proof. Since the dimensions of an L-tetromino are greater than one, we obviously get $n \neq 1$.

We will prove by contradiction that $n \neq 3$.

Assume that the $3 \times m$ rectangle is orientable, i.e., tiled by L-tetrominoes of the same orientation. Let us zoom in to the corners. The top left corner of the $3 \times m$ rectangle (see Figure 19.4) may be covered in three different ways. In cases A and B, we notice that it will be impossible to cover the rest of the rectangle.

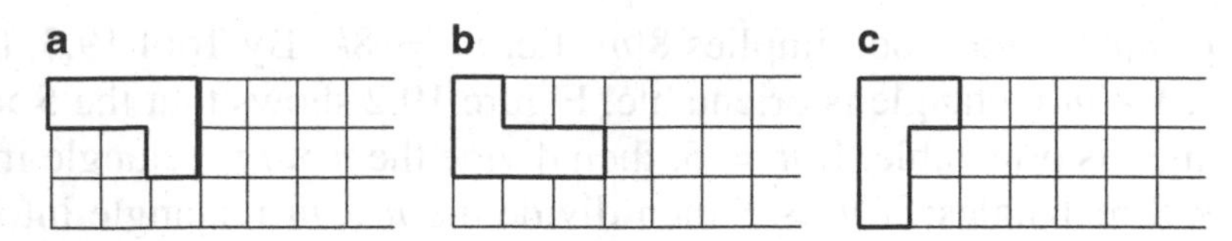

Fig. 19.4

In case C, we can easily establish that the only ways to proceed with the tiling are shown in Figure 19.5:

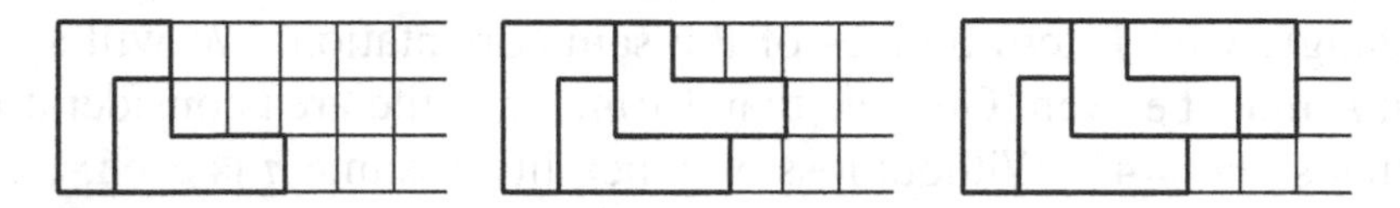

Fig. 19.5

But now the fifth column cannot be completely tiled, which is a contradiction. ■

Now we are ready to summarize Mitya's findings:

Theorem 19.7. *A $n \times m$ rectangle is orientable if and only if* $8|nm$ *and* $n, m \neq 1, 3$.

Proof. Tools 19.3 and 19.4 prove that these conditions are sufficient, while Tools 19.5 and 19.6 prove that the conditions are also necessary. ■

Theorem 19.8. *A rectangle is symmetrically orientable if and only if* $8|nm$ *and* n, m *are both even, or* $16|nm$ *and* $n, m \neq 1, 3$.

Proof. First let us prove sufficiency of the conditions.

If $8|nm$ and n, m are both even, then we can use the same tiling used in Theorem 19.7 (see Figure 19.3).

Now let $16|nm$ and $n, m \neq 1, 3$. Without loss of generality suppose n is odd and $m = 16k$. Divide the $n \times m$ rectangle into two $n \times 8k$ rectangles. Tile one of these $n \times 8k$ rectangles with L-tetrominoes of the same orientation. Now rotate this tiling about the center of the rectangle by 180° — we get a tiling of the other $n \times 8k$ rectangle. The rectangle is tilled in the desired fashion.

Let us now prove the necessity of the conditions. Assume the $n \times m$ rectangle is tiled with L-tetrominoes of the same orientation and that the tiling possesses 2-fold symmetry. By Theorem 19.7, we know $8|nm$ and $n, m \neq 1, 3$. There is only one thing to prove: if n is odd, then $16|m$. This can be done using Tool 19.5.

Pair each L-tetromino with the one that is symmetric to it and let p be the number of such pairs. Color the given rectangle in cyclic row 2-coloring. Since n is odd, it follows that this coloring has 2-fold symmetry, just as our tiling does. Thus, both L-tetrominoes in each pair have the same number of squares of each color. By an argument similar to the one used in Tool 19.5, it follows that $2|p$ implies $16|nm$, which implies $16|m$. ■

In the summer of 2008, Mitya graduated from Columbia University and accepted an invitation to become a Ph.D. student in probability theory at the Courant Institute of Mathematical Sciences at New York University. He promised me, however, not to abandon his excursions into geometry and combinatorics!

Chapter 6
Norton Starr's 3-Dimensional Tromino Tiling

Theorems 2.1 and 2.2 completely described the rectangles in the plane that are tilable by trominoes. We then transferred the problem from the plane to the 3-dimensional space, and solved it in Exercise 6.10. We restate the result here for your convenience.

Theorem 20.1. *(Boltyanski-Soifer) The $m \times n \times k$ parallelepiped can be packed with 3-dimensional L-trominoes if and only if mnk is divisible by 3.*

Where can this train of thought take us? When we tiled a rectangle (Theorem 4.1), we stirred things up by adding a single monomino to our tiling (Problem 4.2). Likewise, here we can add a 3-dimensional monomino or two!

Since we deal with 3-dimensional tilings in this chapter, it should not cause confusion if we write "L-tromino" and "monomino" in place of "3-dimensional L-tromino" and "3-dimensional monomino".

Professor Norton Starr from Amherst College set out to tile cubes with L-trominoes and monominoes (i.e., unit cubes). Of course, it makes the most sense to allow only 1 or 2 monominoes (since 3 monominoes can form a tromino), which by volume consideration reduces our task to cubes of side n where

$$n \equiv 1 \text{ or } 2 \pmod 3.$$

Hence we have two problems:

Problem 20.2. Let $n \equiv 1 \pmod 3$. Can an $n \times n \times n$ cube be tiled with a monomino and L-trominoes? If so, where must the monomino be placed?

A. Soifer, *Geometric Etudes in Combinatorial Mathematics*,
DOI 10.1007/978-0-387-75470-3_6,

Problem 20.3. Let $n \equiv 2 \pmod 3$. Can an $n \times n \times n$ cube be tiled with a pair of monominoes and L-trominoes? If so, where must the pair of monominoes be placed?

Starr solves both problems at once. He uses a simple tool whose proof I hope you will enjoy as much as I do:

Tool 20.4. (Norton Starr [St1]) The $2 \times 2 \times 2$ cube can be tiled with a pair of monominoes and trominoes. The pair of monominoes can be placed in any pair of unit cubes.

Proof. Imagine the $2 \times 2 \times 2$ solid as a lower $2 \times 2 \times 1$ level and an upper $2 \times 2 \times 1$ level. If each level contains one of the monominoes, then on each level a tromino may be placed in the remaining three cells, tiling that level. Both levels and thus packed and the $2 \times 2 \times 2$ solid is thereby tiled.

If both monominoes are in the same level, rotate the $2 \times 2 \times 2$ solid through 90° angle about an edge in the base so that the two monominoes end up in different horizontal levels. Then complete the tiling as above. ■

Norton then creates the following tool:

Tool 20.5. (Norton Starr [St1]) The $4 \times 4 \times 4$ cube **Q** can be tiled with a monomino and trominoes. The monomino can be placed in any unit cube.

Proof. (by Norton Starr with his illustrations, reproduced with the kind permission). It is helpful to envision **Q** partitioned into eight $2 \times 2 \times 2$ subcubes by three slices through the center of **Q** and parallel to its faces.

Rotating **Q** if necessary, we may assume that the monomino denoted by **S**, is in the lower left $2 \times 2 \times 2$ subcube ("LL") at the front of **Q**. Figure 20.1 displays a view of **Q** showing both the subcube LL (containing **S**) and the subcube UL. There are either three or four unoccupied cells in the upper $2 \times 2 \times 1$ half of LL. Place a tromino vertically so that its lower cube fills one of these three or four open cells and its other two cubes lie in the lower half of UL. Now each of the two $2 \times 2 \times 2$ subcubes has two cells occupied. By Tool 20.4, LL and UL may be tiled with trominoes.

There remain six empty $2 \times 2 \times 2$ subcubes in the rest of **Q**. The left image in Figure 20.2 shows an overhead view of **Q** with the top

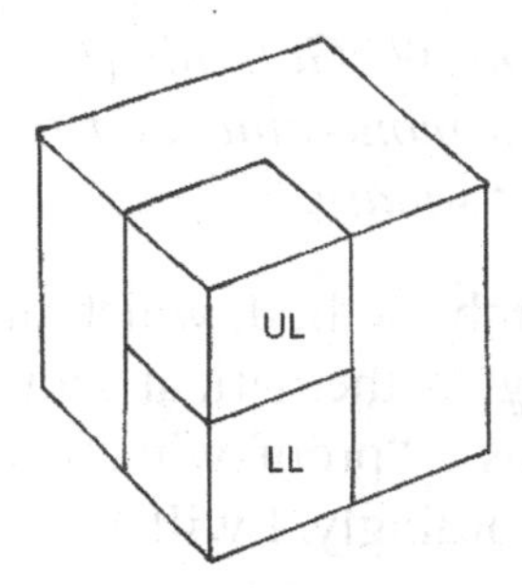

Fig. 20.1

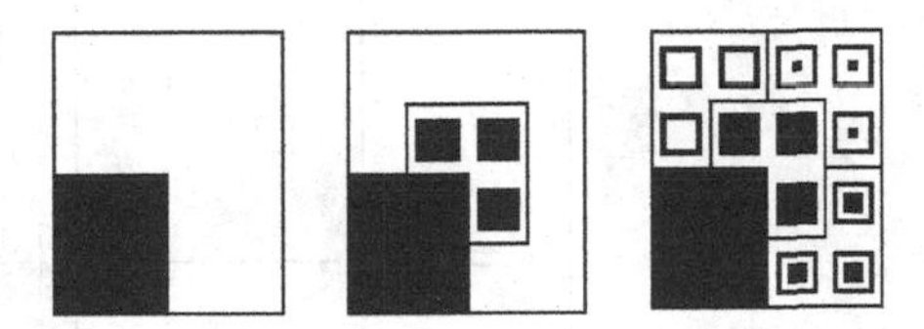

Fig. 20.2

of UL shaded. Note that the unoccupied cells have the L-shape of a large tromino whose three cells are two units on a side. This provides a key to their tiling: using the Golomb method, place four trominoes on top of each other around the center as shown in the middle image of Figure 20.2. The topmost of these four, indicated by ■, has a cube in each of three different $2 \times 2 \times 2$ subcubes in the top half of **Q**. Each of the lower three layers has a tromino in the same location. Indeed, the six previously empty $2 \times 2 \times 2$ subcubes now each have two cells occupied, so they can be tiled by Tool 20.4. (The right image in Figure 20.2 shows how simple it is to complete the tiling in each of these lower three layers, using three trominoes marked □, ⊡, and ▣.)

With the tiling of the each of the six $2 \times 2 \times 2$ subcubes, **Q** is now tiled and Tool 20.5 is proven. ■

In the proof of Tool 20.5, Starr uses "The Golomb Method." What is it?

You may recall meeting Solomon W. Golomb, the inventor of polyomino and tiling problems on square lattices, in Section 1 of Chapter 1. "The Golomb Method" was first used by Golomb to tile a plane. Let us visit it here.

Theorem 20.6. *(Solomon W. Golomb [Go2]) Any plane $2^n \times 2^n$ square can be tiled by a monomino and trominoes. The monomino can be placed in any unit square.*

Proof. We use the Golomb Method, which involves doubling the side of the square and using mathematical induction. Roger B. Nelsen presents this induction as a "proof without words" in [Ne] — see it here in Figure 20.3. Accordingly, I will add no words!

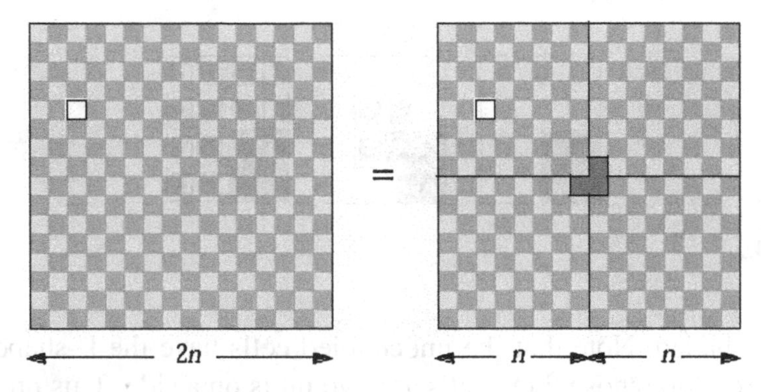

Fig. 20.3 R. B. Nelsen's proof without words [Ne]. Reproduced with the kind permission of the Mathematical Association of America and Roger B. Nelsen.

Starr uses the Golomb Method to solve Problems 20.2 and 20.3 for cubes of side 2^n. He uses Tools 20.4 and 20.5 as the foundation of his inductive argument. "In proving our theorems about cubes **Q** having side length 2^n for $n > 2$, we will reduce the tiling of **Q** to tilings of one or more of its eight subcubes having side length 2^{n-1}," writes Starr [St1].

Theorem 20.7. *(Norton Starr [St1]) Any $2^n \times 2^n \times 2^n$ cube, $n \equiv 1(\text{mod}3)$, can be tiled with a monomino and L-trominoes. Moreover, the monomino can be placed in any unit cube.*

Any $2^n \times 2^n \times 2^n$ cube, $n \equiv 2$ (mod 3), can be tiled with a pair of monominoes and L-trominoes. Moreover, the pair of monominoes can be placed in any pair of unit cubes.

The complete proof of Theorem 20.7 can be found in the October 2008 issue of *Geombinatorics* [St1].

Starr was not deterred by the obvious exponentially sprawling complications of the general case. His proof of the general case is found in [St2].

Theorem 20.8. *(Norton Starr [St2]) Any* $n \times n \times n$ *cube,* $n \equiv 1 (\text{mod } 3)$, *can be tiled with a monomino and L-trominoes. Moreover, the monomino can be placed in any unit cube.*

Any $n \times n \times n$ *cube,* $n \equiv 2 \ (\text{mod } 3)$, *can be tiled with a pair of monominoes and L-trominoes. Moreover, the pair of monominoes can be placed in any pair of unit cubes.*

Starr ends his essay in *Geombinatorics* [St1] with questions about higher dimensional generalization:

Problem 20.9. (Norton Starr [St1]) What are the tromino tiling possibilities for cubes of dimension greater than 3? For a given dimension, does doubling the side length facilitate a recursive argument similar to that used above? How complicated are the proofs for side lengths not equal to 2^n?

Chapter 7
Large Progress in Small Ramsey Numbers

In the first edition [BS] of this book, I wrote that "even $R(4,5)$ or $R(5,5)$ are not known." The problem of calculating $R(4,5)$ has since been settled completely by Brendan D. McKay of the Australian National University and Stanisław P. Radziszowski of the Rochester Institute of Technology.

Ramsey Number Value 21.1. (B. D. McKay and S. P. Radziszowski, [MR4]) $R(4,5) = 25$.

This remarkable result took years of computing to achieve, with the happy end taking place right in front of my eyes. I attended Stanisław Radziszowski's talk in early March of 1993 at the Florida Atlantic University conference. During the talk he mysteriously remarked that the value of $R(4,5)$ might be found very soon. Imagine my surprise when, in a matter of days, I received Staszek's e-mail, announcing the birth of the result seconds after it occurred (this is what computer-aided communication delivers):

From: MX%"spr@cs.rit.edu"
To: ASOIFER
Subj: R(4,5)=25
From: spr@cs.rit.edu (Stanisaw P Radziszowski)
Message-ID: <9303191824.AA22893@rit.cs.rit.edu>
Subject: R(4,5)=25
To: jackkasz@utxvm.cc.utexas.edu,
asoifer@happy.uccs.edu,
goldberg@turing.cs.rpi.edu
Date: Fri, 19 Mar 1993 19-MAR-1993 11:24:29.37 (EST)

DOI 10.1007/978-0-387-75470-3_7,

$R(4,5) = 25$

Brendan D. McKay, Australian National University
Stanisław P. Radziszowski, Rochester Institute of Technology

The Ramsey number R(4, 5) is defined to be the smallest n such that every graph on n vertices has either a clique[3] of order 4 or an independent set[4] of order 5. We have proved that $R(4,5) = 25$. Previously it was only known that R(4,5) is one of the four numbers 25–28. Our proof is computational.

For integers s, t define an (s, t, n)-graph to be an n-vertex graph with no clique of order s or independent set of order t. Suppose that G is a (4, 5, 25)-graph with 25 vertices. If a vertex is removed from G, a (4, 5, 24)-graph H results; moreover, the structure of H can be somewhat restricted by choosing which vertex of G to remove. Our proof consists of constructing all such structure-restricted (4, 5, 24)-graphs and showing that none of them extends to a (4, 5, 25)-graph. In order to reduce the chance of computational error, the entire computation was done in duplicate using independent programs written by each author. The fastest of the two computations required about 3.2 years of cpu time on Sun workstations.

A side result of this computation is a catalogue of 350866 (4, 5, 24)-graphs, which is likely to be most but not all of them.

We wish to thank our institutions for their support. Of particular importance to this work was a grant from the ANU Mathematical Sciences Research Visitors Program.

– bdm@cs.anu.edu.au and spr@cs.rit.edu; March 19, 1993.

Imagine how quickly the amount of computation increases in these "small" Ramsey numbers: "The fastest of the two computations required about 3.2 years of cpu time on Sun Workstation," and "a side result of this computation is a catalogue[5] of 350,866 (4, 5, 24)-graphs"!

---

[3] *Clique* is a subgraph, which is a complete graph, clique of order 4 is a subgraph of K_9.

[4] *Independent set* is a subgraph without any edges.

[5] Later in 1993 this number grew to 350,904.

What about the value of the next Ramsey number, $R(5,5)$? Today's world records in lower and upper bound competitions for the value of $R(5,5)$ are due to Geoffrey Exoo, Brendan McKay and Stanisaw Radziszowskit respectively:

Best Bounds 21.2. ([Ex4][MK5]) $43 \leq R(5,5) \leq 49$.

When the great expert of lower bounds, Exoo, and the great experts of upper bounds McKay and Radziszowski agree that there is evidence for a "strong conjecture," we'd better listen — and record:

McKay-Radziszowski-Exoo's Conjecture 21.3. [MR5] $R(5,5) = 43$.

It may take decades or even a century to settle this number — when we do, we will see whether the three authors of the conjecture are correct. In fact, Paul Erdős liked to explain the difficulties of this problem [Er2]: "It must seem incredible to the uninitiated that in the age of supercomputers $R(5,5)$ is unknown. This, of course, is caused by the so-called combinatorial explosion: there are just too many cases to be checked." He even made up a joke about it, which I heard during his talks in a few different variants:

> Suppose aliens invade the earth and threaten to destroy it in a year if human beings do not find $R(5,5)$. It is, probably, possible to save the earth by putting together the world's best minds and computers. If, however, the invaders were to demand $R(6,6)$, then human beings might as well attempt a preemptive strike without even trying to ponder the problem.

Since 1994, Radziszowski has maintained and revised eleven times a 60-page compendium of "world records" in the sport of small Ramsey numbers [Rad]. This is an invaluable service to the profession. I will present here only the known nontrivial classic 2-color small Ramsey numbers in Table 21.1. Where lower and upper bounds do not coincide, both are listed in the appropriate cell. The cells below the main diagonal are left empty because filling them in would be redundant due to the symmetry of the Ramsey function ($R(m,n) = R(n,m)$; Exercise 9.9). In Table 21.2, I present only a part of Radziszowski's Reference Table for Table 21.1 — you can see the rest in his compendium [Rad] readily available on the Internet. You will find there a

Table 21.1 World Records in Classical 2-color Small Ramsey Numbers

k \ l	3	4	5	6	7	8	9	10	11	12	13	14	15
3	6	9	14	18	23	28	36	40	46	52	59	66	73
								43	51	59	69	78	88
4		18	25	35	49	56	73	92	97	128	133	141	153
				41	61	84	115	149	191	238	291	349	417
5			43	58	80	101	125	143	159	185	209	235	265
			49	87	143	216	316	442		848		1416	
6				102	113	127	169	179	253	262	317		401
				165	298	495	780	1171		2566		5033	
7					205	216	233	289	405	416	511		
					540	1031	1713	2826	4553	6954	10581	15263	22116
8						282	317				817		861
						1870	3583	6090	10630	16944	27490	41525	63620
9							565	580					
							6588	12677	22325	39025	64871	89203	
10								798					
								23556		81200			1265

Table 21.2 References for part of Table 21.1

k \ l	4	5	6	7	8	9	10
3	GG	GG	Kery	Ka2	GR	Ka2	Ex5
				GY	MZ	GR	RK2
4	GG	Ka1	Ex9	Ex3	Ex15	Ex17	HaKr
		MR4	MR5	Mac	Mac	Mac	Mac
5		Ex4	Ex9	CET	HaKr	Ex17	Ex17
		MR5	HZ1	Spe3	Spe3	Mac	Mac
6			Ka1	Ex17	XXR	XXER	Ex17
			Mac	Mac	Mac	Mac	Mac

wealth of other fascinating small Ramsey-related world records, Ramsey numbers defined more broadly than they are here, Ramsey number inequalities, and an extensive bibliography.

Compare Table 21.1 to the tiny Table 9.1 I presented in Section 9 (in fact, Table 9.1 would be even tinier if we remove the trivial row for $r(2,n)$), and you will agree with me that the researchers in small Ramsey numbers have made enormous progress in the past seventeen years. We have a race here: combinatorial explosion versus improvements in computational methods and computers. It seems that computers and mathematicians in this field have gained some!

Chapter 8
The Borsuk Problem Conquered

As we discussed in Section 14 of Chapter 4, Karel Borsuk formulated his celebrated conjecture in 1933:

Borsuk's Conjecture 22.1. For any bounded figure F in R^n, $a(F) \leq n+1$, i.e., F can be decomposed into $n+1$ parts of smaller diameters.

For decades, everyone thought the conjecture was true but no one was able to prove it. Since you likely did not pay too much attention to the forewords, let me quote here from Paul Erdős's foreword to the first edition of this book.

> There is also a more difficult chapter on combinatorial geometry where the famous unsolved conjecture of Borsuk is discussed in great detail. Fifty years ago I spent lots of time trying to prove it. To quote Hardy, I hope younger and stronger hands (or rather brains) will have more success.

I can also see Paul's expectation of the positive answer in his 1946 paper [Er1], when he writes:

> If one could prove that in k-dimensional space the maximum distance [in an n-element set] cannot occur more than kn times, the following conjecture of Borsuk would be established: Each k-dimensional subset of diameter 1 can be decomposed into $k+1$ summands each having diameter <1.

In fact, circa 1940, Paul Erdős thought the Borsuk Conjecture was true and tried to prove it. Forty years later his opinion changed, as we can see in *The Scottish Book* [Er2]:

> Now let me talk about some of the problems which don't seem to be very difficult but still may be of some interest even now, after many years. One of these is a very pretty conjecture by Borsuk

DOI 10.1007/978-0-387-75470-3_8,

which says that if one has a set of diameter 1 in n-dimensional space it can be decomposed into $n + 1$ sets of diameter <1. This is trivial on the line, easy on the plane, difficult in 3-space, and unsolved higher (I suspect that it is false for sufficiently high dimension).

This was a rare dissent; nearly everyone still believed in the validity of the conjecture. One of the active researchers of the Borsuk Conjecture, Andreĭ M. Raigorodskiĭ, lists only two other dissenters [Ra2]:

...only Erdős [Er2], C. A. Rogers [Ro], and D. Larman [L] (as far as we know) dared [to] explicitly express skepticism.[6]

Raigorodskiĭ's list of dissenters is incomplete; in Section 14 of the first edition of *this book*, Boltyanski and Soifer clearly expressed their dissent (actually, we wrote it in June of 1990 — when the first draft of this book was written):

Is there in R^4 a convex polytope M (i.e., a convex hull of a finite point set) such that $a(M) = 6$? Perhaps such a polytope can be found with the help of computers? At any rate, the authors are inclined to believe that a counterexample exists to the Borsuk problem in the case $n \geq 4$.

We did not have to wait long to be proven correct by a striking result: in July of 1993, Jeff Kahn of Rutgers University and Gil Kalai of the Hebrew University published a counterexample to the Borsuk Conjecture in dimension 1326. What followed can only be called a world competition in the Borsukovian sport — new world records have been set for the smallest dimension of the counterexample with lightning speed.

In the same year, 1993, Alon Nilli [N] of Tel Aviv University constructed a counterexample in dimension 946, which was published in 1994. Three years later, in 1997, Andreĭ Raigorodskiĭ, a scientist

[6] This is not true regarding Larman, who does not mention the Borsuk Conjecture at all in [L]. It is absolutely true regarding Rogers. Moreover, Rogers writes with disappointment about his unsuccessful attempt to disprove the Borsuk Conjecture [Ro]: "The results of this note were obtained in an unsuccessful attempt to disprove Borsuk's conjecture. If I felt that the work threw much light on the conjecture, I should be looking at it in this light, rather than putting pen to paper."

at Moscow State University, published a dramatic reduction to 561 of the upper bound [Ra1]. Three more years passed before Bernulf Weißbach of the University of Magdeburg reduced the upper bound by 1 to 560. His work was published in 2000 [We].

On September 27, 2000 Aicke Hinrichs of Jena University submitted a new (impressive) record of 323 [Hi], which was published in 2002. Almost immediately, on February 12, 2002, it was reduced to 321 [Pi] by Oleg Pikhurko, then of Cambridge University and now of Carnegie Mellon University.

Eight days later, on February 20, 2002, Aicke Hinrichs came back with Jena's colleague Christian Richter and set the still reigning record of 298, which was published in 2003. They proved an even stronger result:

Hinrichs-Richter's World Record 22.2. [HR] For $n > 298$, there exists a finite set in the unit sphere in R^n that cannot be partitioned into $n + 11$ sets of smaller diameter.

This still leaves a huge gap and a natural problem:

Open Problem 22.3. Find the smallest $n, 4 \leq n \leq 297$ such that there is a bounded set S in R^n that cannot be partitioned into $n + 1$ sets of smaller diameter.

I thought that such n must be small, not much more than 4, and on May 18, 2008, I shared my thoughts with Hinrichs in an e-mail:

> What do you think is the lowest n for which the conjecture fails? It should be rather small, I think.

On May 28, 2008, Aicke agreed:

> We have the same feeling, I would even dare to conjecture that it is at most ten.

I must record it here:

Conjecture 22.4. (Hinrichs-Richter and Soifer) There exists an $n, 4 \leq n \leq 10$, such that there is a bounded set S in R^n that cannot be partitioned into $n + 1$ sets of smaller diameter.

I would not be surprised if the counterexample lives already in 4-space:

Conjecture 22.5. There is a bounded set S in R^4 that cannot be partitioned into 5 sets of smaller diameter.

On September 13, 2008, Oleg Pikhurko informed me via e-mail that Raigorodskiĭ's intuition is disjoint with mine, for he expects the Borsuk Conjecture to be true for $n = 4$:

> I know that Andreĭ Raigorodskiĭ was trying to prove the conjecture for 4, but apparently he did not succeed.

Roll up your sleeves, open your brains — there is plenty to prove between 4 and 297!

Chapter 9
Etude on the Chromatic Number of the Plane

The Borsuk Conjecture survived for sixty-one years from 1932 to its negative resolution in 1993. The problem I am going to share with you here is now sixty years old. It may require another century, may be centuries, to be settled. This is my favorite unsolved problem in all of mathematics, because it is so easy to understand — I have presented it to middle school audiences — but very hard, if it is even possible, to solve. Not only to solve, but even to narrow down the range in the general case is difficult! Here is the problem:

Chromatic Number of the Plane Problem 23.1. (Edward Nelson, 1950) Find the smallest number of colors sufficient for coloring the planc in such a way that no two points of the same color are unit distance apart.

This number is called the *chromatic number of the plane* and is denoted χ. *To color the plane* means to assign one color to every point of the plane. Please note that we color without any restrictions and are not limited to "nice," tiling-like or map-like colorings. Given a positive integer n, we say that the plane is *n-colored* if every point of the plane is assigned one of the given n colors.

A *segment* here refers to a 2-point set. Similarly, a *polygon* refers to a finite set of points. A *monochromatic set* is a set whose elements are all assigned the same color.

This is a principally modern problem: I do not think that ancient geometers asked this kind of questions. What are the bounds of the chromatic number χ of the plane? At first glance it does

A. Soifer, *Geometric Etudes in Combinatorial Mathematics*,
DOI 10.1007/978-0-387-75470-3_9,

not seem obvious that χ is even finite. It is, as the following simple construction shows:

Upper Bound 23.2. (John Isbell, 1950) There is a 7-coloring of the plane that contains no monochromatic unit length segments, i.e.,

$$\chi \leq 7.$$

Proof. Tile the plane with regular hexagons of side 1. Color one hexagon in color 1 and its six neighbors in colors 2, 3, ..., 7 (Figure 23.1). The union of these seven hexagons forms a highly symmetric flower-like polygon P of 18 sides. Translates of P tile the plane and determine the desired 7-coloring of the plane. It is easy to compute (do) that each color does not have any monochromatic segments of length d for $2 < d < \sqrt{7}$. Thus, if we shrink all linear sizes by a factor of, say, 2.1, we will get a 7-coloring of the plane that forbids monochromatic unit distance segments. (Observe that due to the above inequality, we have enough cushion so that it does not matter which of the two adjacent colors we use to color the boundaries of hexagons). ■

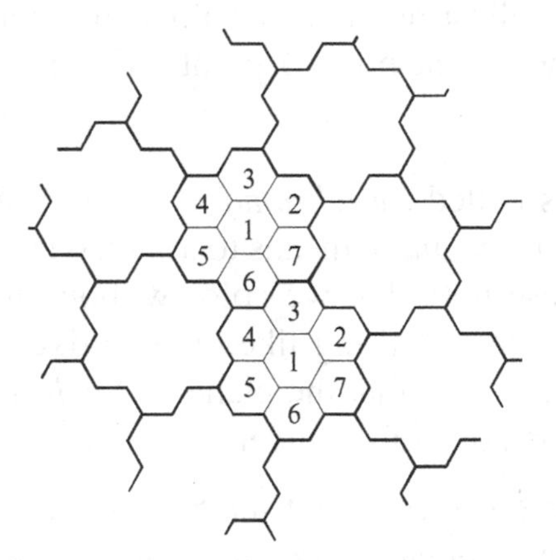

Fig. 23.1 A 7-coloring of the plane forbidding monochromatic distance 1

The creator of the problem was also first to find a lower bound:

Lower Bound 23.3. (Edward Nelson, 1950) No matter how the plane is 3-colored, it contains a monochromatic unit length segment, i.e.,

$$\chi \geq 4.$$

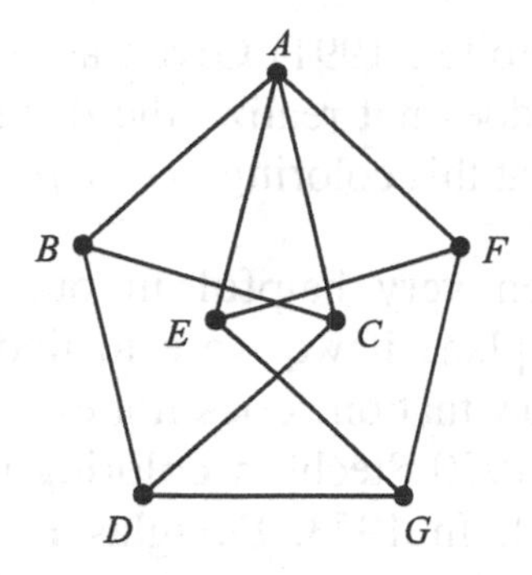

Fig. 23.2 The Mosers Spindle

Proof. (by the Canadian geometers Leo and William Moser, 1961, [MM]). Toss on the given 3-colored plane what we now call the *Mosers Spindle* (Figure 23.2). Every edge in the spindle has length 1.

Assume that the seven vertices of the spindle do not contain a monochromatic unit length segment. Suppose the colors used to color the plane are red, white, and blue. The solution now follows the children's song: "*A, B, C, D, E, F, G, ...*".

Let point A be red; then B and C must be white and blue. Therefore D is red. Similarly E and F must be white and blue, so G is red. We got a monochromatic segment DG of length 1, contradicting our assumption.

Observe: the Mosers Spindle worked for us in solving Problem 23.3 precisely because *any* three points of the spindle contain two points distance 1 apart. This implies that *at most two points can be of the same color in a Mosers spindle that forbids monochromatic distance 1.* ■

It is amazing that the easily proven bounds 23.2 and 23.3 give us the best known bounds for the chromatic number of the plane. The problem and these bounds date back to 1950 (read an exciting investigation into the history of this problem in *The Mathematical Coloring Book* [S7]). Still, all we know is that $\chi = 4$, or 5, or 6, or 7.

A very broad range! Which do you think is the exact value of χ?

In 1991, I waged an all out attack on the upper bound. I tried to reduce 7 to 6 and failed. But in the process I discovered a continuum of interesting 6-colorings of the plane. In order to assess and compare various colorings, we can make use of the definition of the *type of coloring* which I published in 1992 [S4].

Definition 23.4. (A. Soifer, 1991) Given an n-coloring of the plane such that the color i does not realize the distance d_i $(1 \leq i \leq n)$. Then we would say that this coloring is of *type* $(d_1, d_2, \ldots, d_n)$.

It would have been very helpful in our search for the chromatic number of the plane if we were to find a 6-coloring of type (1,1,1,1,1,1), or to show that one does not exist. With the appropriate choice of a unit, the 1970 Stechkin coloring that appeared in [Rai] had type $(1,1,1,1,\frac{1}{2},\frac{1}{2})$. In 1973, Douglas R. Woodall [Wo] found the second 6-coloring of the plane with not all distances realized in any color. Woodall's coloring had a special property that he desired for his purposes: each of the six monochromatic sets was closed. His example, however, had three "missing distances:" it had type $(1,1,1,\frac{1}{\sqrt{3}},\frac{1}{\sqrt{3}},\frac{1}{2\sqrt{3}})$. Woodall unsuccessfully tried to reduce the number of distinct distances. As he wrote, "I have not managed to make two of the three 'missing distances' equal in this way ([Wo], p. 193)."

In my first example of a 6-coloring, I equalized *almost* all distances:

Six-Coloring 23.5. (A. Soifer [S4]) There is a 6-coloring of the plane of type $(1, 1, 1, 1, 1, \frac{1}{\sqrt{5}})$.

Proof. We start with two squares, one of side 2 and the other of diagonal 1 (Figure 23.3). We can use them to create a tiling of the plane with squares and (non-regular) octagons (Figure 23.5). Colors $1, \ldots,$ 5 will color the octagons; we will color all squares in color 6. With each octagon and each square we include half of its boundary (bold lines in Figure 23.4) without the endpoints. It is easy to verify (do) that the distance $\sqrt{5}$ is not realized by any of the colors $1, \ldots, 5$; and

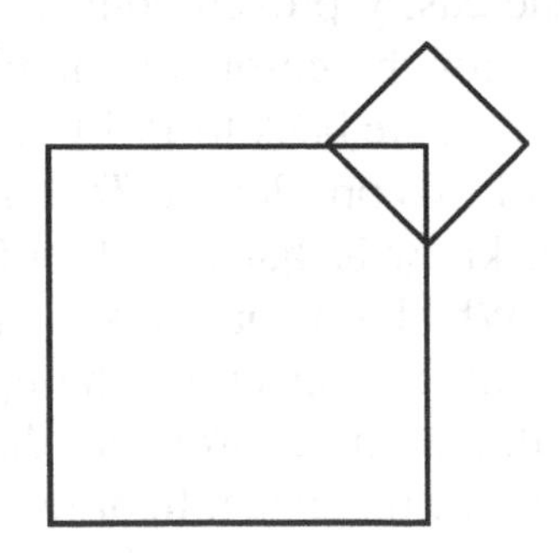

Fig. 23.3

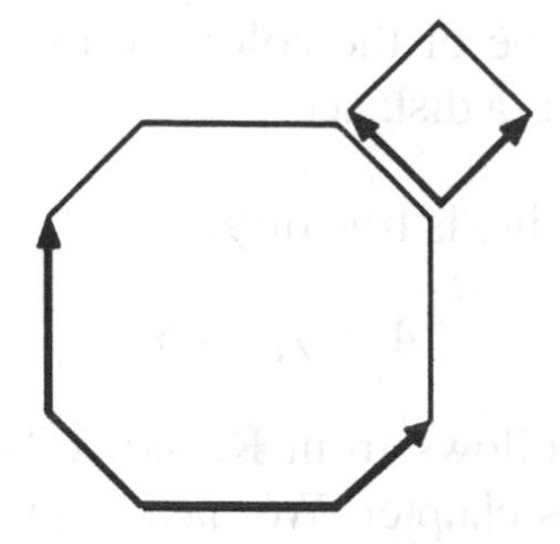

Fig. 23.4

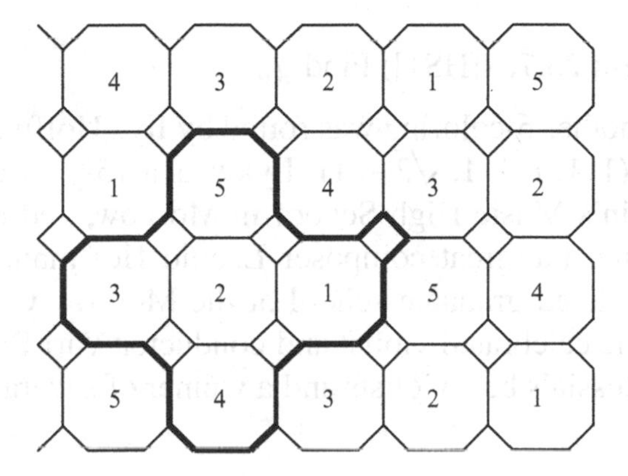

Fig. 23.5 Soifer's 6-Coloring of the Plane

the distance 1 is not realized by the color 6. By shrinking all linear sizes by a factor of $\sqrt{5}$, we get the 6-coloring of type $(1,1,1,1,1,\frac{1}{\sqrt{5}})$.

To verify the latter, observe the unit of the construction that is bounded by the bold line in Figure 23.5 and consists of five octagons and five squares; its translates tile the plane. ■

I had mixed feelings when I obtained result 23.5 in August of 1991. On the one hand, I knew the result was close but no cigar; after all, a 6-coloring of type (1,1,1,1,1,1) still had not been found. On the other hand, I thought that the latter 6-coloring might not exist, and if so, my 6-coloring would be best possible. This short paper [S4] gave birth to a new definition and an open problem.

Definition 23.6. ([HS1]) *The almost chromatic number* χ_a *of the plane* is the minimal number of colors that are required to color the plane so that all but one of the colors forbid unit distance, and the remaining color forbids a distance.

We know the following is true of χ_a:

$$4 \leq \chi_a \leq 6.$$

The lower bound follows from Raiskii's Theorem 23.12 that we will prove later in this chapter. We have proven the upper bound in result 23.5 above [S4]. This naturally gave birth to a new problem, which is still open:

Open Problem 23.7. ([HS1]) Find χ_a.

In 1993, another 6-coloring was found by Ilya Hoffman and I [HS]. Its type was $(1, 1, 1, 1, 1, \sqrt{2} - 1)$. Ilya was a 15-year old violin student at Gnesin's Music High School in Moscow, and the son of my first cousin and the great composer Leonid Hoffman. Today at 31, Ilya has completed graduate school at the Moscow Conservatory in the class of the celebrated violist and conductor Yuri Bashmet, and is now one of Russia's best violists and a winner of several international competitions.

Second Six-Coloring 23.8. (I. Hoffman and A. Soifer [HS]) There is a 6-coloring of the plane of type $(1, 1, 1, 1, 1, \sqrt{2} - 1)$.

Proof. We tile the plane with squares of diagonals 1 and $\sqrt{2} - 1$ (Figure 23.6). We use colors $1, \ldots, 5$ for the larger squares and color 6 for the smaller squares. With each square we include half of its boundary, the left and lower sides, without the endpoints (Figure 23.7).

To verify that this coloring does the job, observe the unit of the construction that is bounded by the bold line in Figure 23.6 and consists of 5 large and 5 small squares; its translates tile the plane. ∎

Examples 23.5 and 23.8 prompted me to introduce a new terminology for this problem, and to translate the results and problems into this new language.

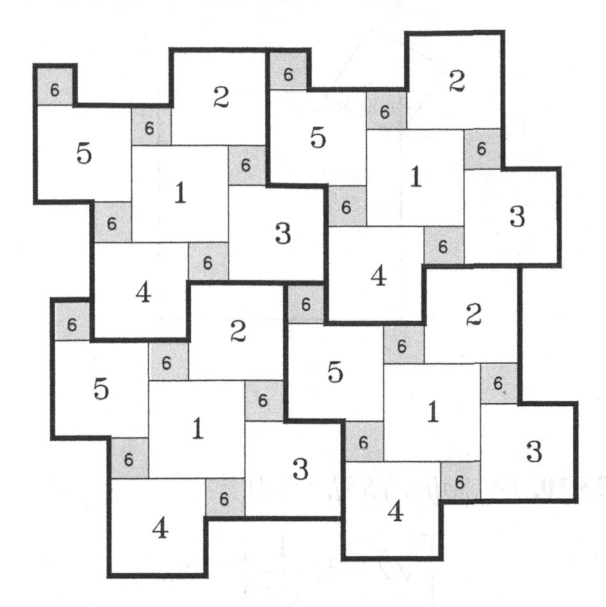

Fig. 23.6 Hoffman-Soifer's 6-Coloring of the Plane

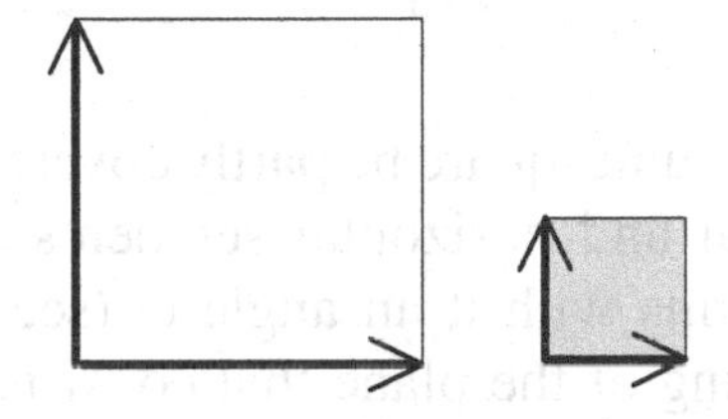

Fig. 23.7

Open Problem 23.9. (A. Soifer [S5], [S6]) Find the *6-realizable set* X_6 of all positive numbers α such that there exists a 6-coloring of the plane of type $(1, 1, 1, 1, 1, \alpha)$.

In this new language, Six-Colorings 23.5 and 23.8 can be written as follows:

$$\frac{1}{\sqrt{5}}, \sqrt{2} - 1 \in \mathrm{X}_6.$$

Now we have two examples of "working" 6-colorings. But what do they have in common? After a while I realized that they were two extreme examples of the general case, a continuum of "working" 6-colorings!

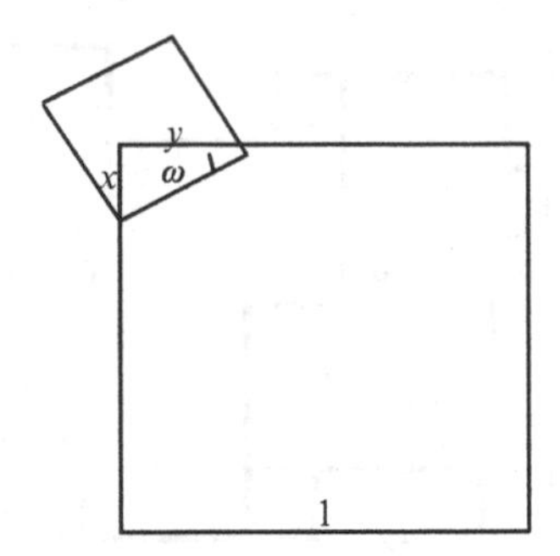

Fig. 23.8

Theorem 23.10. *(A. Soifer [S5], [S6])*

$$\left[\sqrt{2}-1, \frac{1}{\sqrt{5}}\right] \subseteq X_6,$$

i.e., for every $\alpha \in \left[\sqrt{2}-1, \frac{1}{\sqrt{5}}\right]$ *there is a 6-coloring of the plane of type* $(1, 1, 1, 1, 1, \alpha)$.[7]

Sketch of Proof. Let a unit square be partly covered by a smaller square that cuts off vertical and horizontal segments of lengths x and y, respectively, and forms with it an angle ω (see Figure 23.9). These squares induce a tiling of the plane that consists of non-regular congruent octagons and "small" squares (Figure 23.9).

Now we are ready to color the tiling in 6 colors. Denote the unit of our construction bounded by a bold line (Figure 23.10) and consisting of 5 octagons and 5 "small" squares by F. Use colors 1 through 5 for the octagons inside F and color 6 for all "small" squares. Include in the coloring the part of each boundary shown in bold in Figure 23.9. Translates of F tile the plane and thus determine a 6-coloring of the plane.

We now select the sizes of the two squares and parameters x, y, and ω to guarantee that each octagonal color forbids distance 1, and the color of "small" squares forbids given in advance distance α from the allowed segment. Read proof of existence in [S6] and [S7]. ■

[7] $[a, b]$, $a < b$, as usual, stands for the line segment including its endpoints a and b.

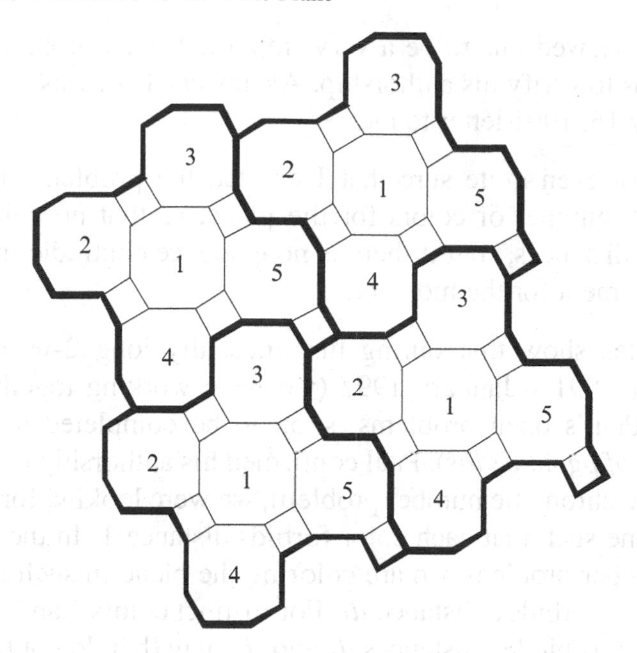

Fig. 23.9 Soifer's continuum of 6-colorings of the plane

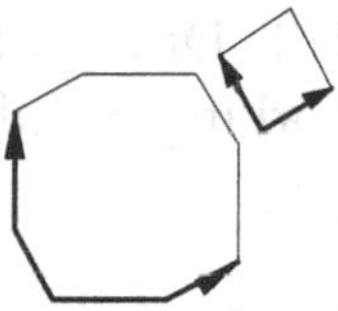

Fig. 23.10

Paul Erdős asked whether it was true that if the plane is partitioned (colored) into three disjoint subsets, one of the subsets must realize all distances. Soon the problem took on its current "appearance." Here it is:

Erdős's Open Problem 23.11. What is the smallest number of colors needed to color the plane so that no color realizes all distances?

This number had to have a name, and so in 1992 [S5]. I named it the *polychromatic number of the plane* and denoted it χ_p. The name and the notation seemed so natural, that by now it has become standard, and has (without credit) appeared in such encyclopedic books as [JT] and [GO].

Since I viewed this to be a very important open problem, I asked Paul Erdős to verify his authorship. As always, Paul was very modest in his July 16, 1991 letter to me:

> I am not even quite sure that I created the problem "find the smallest number of colors for the plane, so that no color realizes all distances," but if there is no evidence contradicting it we can assume it for the moment.

My notes show that during his unusually long 2-week visit in December 1991 – January 1992 (we were working together on the book of Paul's open problems, soon to be completed and entitled *Problems of pgom Erdős*), Paul confirmed his authorship of this problem. In the chromatic number problem, we were looking for colorings of the plane such that each color forbids distance 1. In the polychromatic number problem, we are coloring the plane in such a way that each color i forbids a distance d_i. For distinct colors i and j, the corresponding forbidden distances d_i and d_j may (but do not have to) be distinct.

Nothing had been discovered during the first twelve years of this problem's life. Then in 1970, Dmitry E. Raiskii, a student of the Moscow High School for Working Youth[8] 105, published ([Rai]) the lower and upper bounds for χ_p.

Raiskii's Theorem 23.12. (D. E. Raiskii [Rai]) $4 \le \chi_p$.

In 2003, the Russian-turned-Israeli mathematician Alexei Kanel-Belov communicated to me an incredibly beautiful short proof of Raiskii's Theorem by his student. The proof was found by Alexei Merkov, a 10th grader from Moscow High School 91.

Proof of the Raiskii's Theorem. (A. Merkov) Assume the plane is colored in three colors, red, white and blue, but each color forbids distance r, w, and b, respectively. Equip the 3-colored plane with the Cartesian coordinates and construct in the plane three 7-point sets S_r, S_w, and S_b, each being a Mosers Spindle (see Figure 23.2). Position such that all S_r, S_w, and S_b share the origin O as one of their seven vertices and have edges three equal to r, w, and b, respectively. This construction defines six "red" vectors $v_1, \ldots, v_6$ from O to each point

[8] Students in such high schools hold regular jobs during the day and attend classes at night.

of S_r; six "white" vectors $v_7, ..., v_{12}$ from O to the points of S_w; and six "blue" vectors $v_{13}, ..., v_{18}$ from O to the points of S_b: eighteen vectors in all.

Introduce now the 18-dimensional Euclidean space $\mathbb{R}^{18}$ and a function M from $\mathbb{R}^{18}$ to the plane $\mathbb{R}^2$ naturally defined as follows: $(a_1, ..., a_{18}) \mapsto a_1v_1 + ... + a_{18}v_{18}$. This function induces a 3-coloring of $\mathbb{R}^{18}$ by assigning a point of $\mathbb{R}^{18}$ the color of the corresponding point of $\mathbb{R}^2$. The first six axes of $\mathbb{R}^{18}$ we will call "red", the next six axes "white", and the last six axes "blue."

Define the subset of $\mathbb{R}^{18}$ of all points whose coordinates include at most one coordinate equal to 1 for each of the three colors of the axes, and the rest 0 by W. It is easy to verify (do) that W consists of 7^3 points. For any fixed array of coordinates allowable in W on white and blue axes, we get the 7-element set A of points in W having these fixed coordinates. The image $M(A)$ of the set A under the map M forms in the plane a translation of the original 7-point set S_r. If we fix another array of white and blue coordinates, we get another 7-element set in $\mathbb{R}^{18}$, whose image under M forms in $\mathbb{R}^2$ another translation of S_r. Thus, W is partitioned into 7^2 subsets, each of which maps into a translate of S_r.

Now recall the observation we made in the last paragraph of the proof of the lower bound 23.3 It implies that any translate of the Mosers Spindle S_r contains at most two red points out of its seven points. Since W has been *partitioned* into translates of S_r, at most $\frac{2}{7}$ of the points of W are red. We can start over again, and in a similar way show that at most $\frac{2}{7}$ of the points of W are white, and similarly to show that at most $\frac{2}{7}$ of the points of W are blue. But $\frac{2}{7} + \frac{2}{7} + \frac{2}{7}$ does not add up to 1! This contradiction implies that at least one of the colors realizes all distances, as required. ■

Coloring 23.5 and Raiskii's Theorem 23.12 prove that $4 \le \chi_p \le 6$. These are still the best bounds known today.

For 60 years, mathematics has witnessed a variety of approaches used in attempts to settle the chromatic number of the plane. Tools from graph theory, topology, measure theory, abstract algebra, discrete and combinatorial geometry have been tried — and yet no improvement has been attained in the general case. The range of the chromatic number of the plane remains as wide open as ever: $\chi = 4, 5, 6$, or 7.

I felt — and wrote a number of years ago — that such a wide range was an embarrassment for mathematicians. After 60 years of very active work on the problem, we have not even been able to confidently conjecture the answer. Have mathematicians been inept, or is the problem that good? Have we been missing something in our assault on the problem?

In October 2002 at Rutgers University in Piscataway, New Jersey, I joined forces with Saharon Shelah, a genius of problem solving, for a week of joint research. As a result, we showed that the chromatic number of a graph, and in particular of the plane, may depend upon the system of axioms we choose for set theory. This may explain why this problem is so hard to solve, and even narrow down the 4-to-7 range. Read a detailed report about this work and much, much more about the mathematics of coloring and its exciting history in my new *Mathematical Coloring Book* [S7].

Farewell to the Reader

Thank you for holding my book in your hands. I welcome your ideas, comments, conjectures, solutions of problems presented here and new problems you may create. They may inspire a new edition of this book. I hope we will meet again on the pages of my other books.

As Paul Erdős used to say at the end of his lectures, "everything comes to an end," and so has this book. However, if you are inclined to continue your explorations of mathematics with me, I have good news for you. This book is one of my 8 books that Springer has or soon is going to publish.

If you are receptive to a visual appeal of geometry, you may wish to read the new expanded edition of *How Does One Cut a Triangle?* [S10]. Its first edition [S2] was published in 1990. Furthermore, these books offers a fragment of "real" mathematics, a demonstration of synthesis, where ideas from various branches of mathematics – geometry, trigonometry, linear algebra, extensions of rings, – work together to produce a geometric result.

If you are interested in a mixture of exciting problems, I recommend you to work through the new edition of the book *Mathematics as Problem Solving* [S9]. Its first edition [S1] came out in 1987.

You will find an even greater variety of problems in the new book *Colorado Mathematical Olympiad: The First 20 Years and Further Explorations* [S11]. The book of *The First 10 Years* was published in 1994 [S8]. Moreover, these books also offer 20 "bridges" from problems of mathematical Olympiads to problems of "real" mathematics. You will even find there open problems that could inspire you to start mathematical research on your own!

The Election Day, November 4, 2008 (the "yes-we-can" day) saw the release of the book I dreamed of and worked on for 18 year, *The Mathematical Coloring Book: Mathematics of Coloring and the Colorful Life of Its Creators* [S7]. This voluminous book offers a beautiful mathematics of coloring (so-called *Ramsey Theory*), historical investigations into lives of mathematicians, from the Nazi time in Germany to the devastated by World War II Netherlands. The history allowed me to pose questions which have not lost its urgency today, such as the role of a scholar in tyranny. The book presents aesthetics of mathematics as an art, philosophy of its foundations, and psychology of mathematical and historical discovery. The Nobel Laureate Boris Pasternak expressed my goals in this book better and more concisely than I could – great poets often do it well:

I bring here all: what have I lived thru,
And that what keeps my soul alive,
My rectitude and aspirations,
And what have seen my own eyes.[10]

My next book will not include mathematics. However, the great XX century mathematician will be the hero of the book, which will therefore be entitled *Life and Fate: In Search of Van der Waerden* [S12]. I hope it will be published in 2011.

The book of open problems of the legendary mathematician Paul Erdős will come next, likely in 2012: *Problems of pgom Erdős* [ErS]. I would not have attempted to write it, but in 1990 Paul asked me to join him in this endeavor, and thus it will be our joint book. As you may know, Paul Erdős (1913–1996) was the greatest problem creator of all time. You will be able to work on his problems because no knowledge is required to understand the majority of them. Moreover, many problems allow young mathematicians to advance, find partial solutions.

The book after Erdős would be either *The Art on the Frontier of Cultures: The Fang People of West Equatorial Africa and Their Neighbors*, or *Memory in Flashback*. The former would be a result of my ongoing study of African Art and culture, inspired by the great anthropologist, my hero and friend James W. Fernandez. The latter will be a collection of humorous and noteworthy moments of my life, meetings

[10] [Pas], Translated especially for *The Mathematical Coloring Book* [S7] by Ilya Hoffman.

with great people of many creative professions, and lessons from both sides of the Atlantic.

Having finished this book, you have become my alumnus, the title that carries responsibility to stay in touch, to send me your most enjoyable solutions, your new problems, conjectures, suggestions and ideas. Rest assured: I will always be delighted to hear back from you!

References

[B1] Boltyanski, V. G., A problem about the illumination of the boundary of a convex body. *Izv. Moldavsk Filiala Akad. Nauk SSSR* 10(76) (1960), 79–86 (Russian).

[B2] Boltyanski, V. G., O razbienii ploskikh figur na chasti men'shevo diametra (On partitioning plane figures into parts of smaller diameters). *Colloquium Math. Warszawa* 21(2) (1970), 253–263 (Russian).

[B3] Boltyanski, V. G., Generalization of a certain theorem of Szökefalvi-Nagy, *Dokl. Adad. Nauk SSSR* 228(2) (1976), 265–268 (Russian).

[B4] Boltyanski, V. G., Several theorems of combinatorial geometry. *Mat. Zametki* 21(1) (1977), 117–124 (Russian).

[B5] Boltyanski, V. G., and Soltan, P. S., Combinatorial geometry and convexity classes. *Uspehi Mat. Nauk* 33, 1(199) (1978), 3–42, 262 (Russian).

[B6] Boltyanski, V. G., A new step in the solution of the Szökefalvi-Nagy problem. *Disc. and Comput. Geom.* 8(1) (1992), 27–49.

[Bo1] Borsuk, K., Über die Zerlegung einer Euklidischen n-dimensionalen Vollkugel in n Mengen. *Verh. Internat. Math. Kongr.* 2 (1932), 192 (German).

[Bo2] Borsuk, K., Drei Sätze über die n-dimensionale Sphäre. *Fund. Math.* 20 (1933), 177–190 (German).

[BG] Boltyanski, V.G., and I. Ts. Gohberg, Results and Problems in Combinatorial Geometry, Cambridge University Press, Cambridge, 1985.

[BMP] Brass, P., Moser, W., and Pach, J., *Research problems in discrete geomtery*, Springer, New York, 2005.

[BS] Boltyanski, V. G., and Soifer, A., Geometric etudes in combinatorial mathematics, Center for Excellence in Mathematical Education, Colorado Springs, 1991.

[CET] Calkin, N. J., Erdős, P., and Tovey, C. A., New Ramsey Bounds from Cyclic Graphs of Prime Order. *SIAM J. Discrete Math.* 10 (1997), 381–387.

[DGK] Danzer, L., Grünbaum, B., and Klee, V., Helly's theorem and its relatives. *Proc. Symp. Pure Math.* 7 (Convexity), (1963), 101–180.

[El] Eggleston, H. G., Covering a three-dimensional set with sets of smaller diameter. *London Math. Soc.* 30(1) (1955), 11–24.

[E2] Eggleston, H. G., *Convexity.* Cambridge University Press, London, 1958.

[Er1] Erdős, P., On Sets of Distances on n Points, *Amer. Math. Monthly* 53(5) (1946), 248–250.

[Er2] Erdős, P., My Scottish Book "Problems." *The Scottish Book: Mathematics from the Scottish Café*, R. D. Mauldin (ed.), Birkhäuser, Boston, 1981.

[Er3] Erdős, P., Problems and results in discrete mathematics, Trends in discrete mathematics, *Discrete Math.* 136 (1–3) (1994), 53–73.

[ErS] Erdős, P., and Soifer, A., *Problems of pgom Erdős*, Springer, New York, 2012, to appear.

[Ex1] Exoo, G., Applying Optimization Algorithm to Ramsey Problems. *Graph Theory, Combinatorics, Algorithms, and Applications* Y. Alavi (ed.), SIAM Philadelphia, 1989, 175–179.

[Ex2] Exoo, G., A Lower Bound for $R(5, 5)$, *J. Graph Theory* 13 (1989), 97–98.

[Ex3] Exoo, G., On Two Classical Ramsey Numbers of the Form $R\ (3, n)$, *SIAM J. DiscreteMath* 2 (1989), 488–490.

[Ex4] Exoo, G., Announcement: On the Ramsey Numbers $R(4,6)$, $R(5,6)$ and $R(3,12)$, *Ars Combinatorica* 35 (1993), 85. (The construction of a graph proving $R(4,6) \geqslant 35$ is presented in detail at http://ginger.indstate.edu/ge/RAMSEY (2001)).

[Ex5] Exoo, G., Some Applications of *pq*-groups in Graph Theory. *Discussiones Mathematicae Graph Theory* 24 (2004), 109–114. Constructions available at http://ginger.indstate.edu/ge/RAMSEY.

[Ex6] Exoo, G., Personal communication to S. P. Radziszowski (2005-2006). (Constructions at http://ginger.indstate.edu/ge/RAMSEY).

[F] Fáry, I., On straight line representations of planar graphs. *Acta Sci. Math.* 11 (1948), 229–233.

[Ga] Gale, D., On inscribing n-dimensional sets in a regular n-simplex. *Proc. Amer. Math. Soc.* 4 (1953), 222–225.

[Go] Goodman, J. E., and O'Rourke, J., *Handbook of discrete and computational geometry*, CRC Press, Boca Raton, 1997.

[Go1] Golomb, S.W., Checkerboards and Polyominoes. *Amer. Math. Monthly* 61(10) (1954), 672–682.

[Go2] Golomb, S.W., *Polyominoes,* Charles Scribner and Sons, New York, 1965.

[Go3] Golomb, S.W., *Polyominoes*, 2nd ed., Princeton University Press, Princeton, 1994.

[GK] Golomb, S.W., Klarner, D.A., Covering a rectangle with L-tetrominoes. *Amer. Math Monthly* 70(7) (1963), 760–761.

[GY] Graver, J. E., and Yackel, J., Some Graph Theoretic Results Associated with Ramsey's Theorem. *J. Combin. Theory* 4 (1968), 125–175.

[GR] Grinstead, C., and Roberts, S., On the Ramsey Numbers $R\ (3,8)$ and $R\ (3,9)$, *J. Combin.Theory Ser. B* 33 (1982), 27–51.

[G1] Grünbaum, B., A simple proof of Borsuk's conjecture in three dimensions. *Proc. Cambridge Philos. Soc.* 53 (1957), 776–778.

[G2] Grünbaum, B., *Convex Polytopes*, John Wiley & Sons, New York, 1967.

[GS] Grünbaum, B. and Shephard, G. C., *Tilings and Patterns,* W. H. Freeman and Co., New York, 1987.

[Gu] Guy, R. K., Crossing number of graphs. *Graph Theory and Applications.* Springer-Verlag, New York, 1972, 111–124.

[H1] Hadwiger, H., Überdeckung einer Menge durch Mengen kleineren Durchsmessers. *Comm. Math. Helv.* 18 (1945–46), 73–75 (German).

[H2] Hadwiger, H., Überdeckung einer Menge durch Mengen kleineren Durchsmessers. *Comm. Math. Helv.* 19 (1946–47), 72–73 (German).

[H3] Hadwiger, H., *Altes und neues über konvexe Körper*, Birkhäuser, Basel-Stuttgart, 1955 (German).

[H4] Hadwiger, H., Ungelöste Probleme. *Elem. der Math.* 12(20) (1957) 121 (German).

[HDK] Hadwiger, H., Debrunner, H., and Klee, V., *Combinatorial Geometry in the Plane*, Holt, Rinehart and Winston, New York, 1964.

[HRW] Harary, F., Robinson, R. W., and Wormald, N., Isomorphic Factorizations I: Complete Graphs. *Trans. Amer. Math. Soc.*, 242 (1978), 243–260.

[Har] Hardy, G. H., *A Mathematician's Apology*, Cambridge University Press, Cambridge, 1940.

[HaKr] Harborth, H., and Krause, S., Ramsey Numbers for Circulant Colorings. *Congressus Numerantium*, 161 (2003), 139–150.

[He1] Helly, E., Über Mengen konvexer Körper mit gemeinschaftlichen Punkten. *Jber. Deutsch. Math. Verein.* 32 (1923), 175–176 (German).

[He2] Helly, E., Über Systeme von abgeschlossenen Mengen mit gemeinschaftlichen Punkten. *Monatsh. Math.* 37 (1930), 281–302 (German).

[Hep] Heppes, A., Terbeli ponthalmazok felosztása kisseb atmeröjü reszhalmazok összegere, *Mat. es fiz tud közl.* 7 (1957), 413–416 (Hungarian).

[Hi] Hinrichs, A., Spherical codes and Borsuk's conjecture. *Discrete Math.* 243 (2002), 253–256.

[HR] Hinrichs, A., and Richter, C., New sets with large Borsuk numbers. *Discrete Math.* 270 (2003), 137–147.

[HS] Hoffman, I., and Soifer, A., Almost chromatic number of the plane. *Geombinatorics* III(2) (1993), 38–40.

[HS1] Hoffman, I., and Soifer, A., Almost chromatic number of the plane. *Geombinatorics* III(2) (1993), 38–40.

[HZ] Huang, Y. R., and Zhang, K. M., An New Upper Bound Formula for Two Color Classical Ramsey Numbers. *J. Combin. Math. and Combin. Computing* 28 (1998), 347–350.

[JT] Jensen, T. R., and Toft, B., *Graph coloring problems*, Wiley, New York, 1995.

[KK] Kahn, J., and Kalai, G., A counterexample to Borsuk's conjecture. *Bull. Amer. Math. Soc.* 29 (1993), 60–62.

[Ka1] Kalbfleisch, J. G., Construction of Special Edge-Chromatic Graphs. *Canad. Math. Bull.* 8 (1965), 575–584.

[Ka2] Kalbfleisch, J. G., *Chromatic Graphs and Ramsey's Theorem*, Ph.D. thesis, University of Waterloo, January 1966.

[Kar] Karabash, D., Small Extension to *Geometric Etudes*, *Geombinatorics* XVI(2) (2006), 262–265.

[Kéry] Kéry, G., On a Theorem of Ramsey. *Matematikai Lapok* 15 (1964), 204–224 (Hungarian).

[K] Kharazishvili, A. B., K zadache osvetshenia (On the problem of illumination). *Soob. AN GSSR* (1973), 289–291 (Russian).

[Ki] Kincses, J., The classification of 3- and 4-Helly dimensional convex bodies. *Geometriae Dedicata* 22 (1987), 283–301.

[Kl] Klarner, D. A., Packing a rectangle with congruent N-ominoes. *J. Combin. Theory* 7 (1969), 107–115.

[Ko] Kolodziejczyk, K., Borsuk's covering and planar sets with unique completion (to appear).

[Ku] Kuratowski, K., Sur le probleme des courbes gauches en topologie. *Fund. Math.* 16 (1930), 271–283 (French).

[L] Larman, D., Open problem 6. *Convexity and Graph Theory*, M. Rozenfeld and J. Zaks (eds.), *Ann. Discrete Math.* (20) (1984), 336.

[LS] Liusternik, L. A., and Shnirelman, L. G., *Topologicheskije Metody v Variazionnyh Zadachah* (*Topological Methods in Variational Problems*), Moscow, (1930), (Russian).

[Mac] Mackey, J., Combinatorial Remedies, Ph.D. Thesis, University of Hawaii, 1994.

[MR1] McKay, B. D., and Radziszowski, S. P., $R(4, 5) = 25$. *J. Graph Theory* 19 (1995), 309–322.

[MR2] McKay, B. D., and Radziszowski, S. P., Subgraph Counting Identities and Ramsey Numbers. *J. Combin. Theory Ser. B* 69 (1997), 193–209.

[MR3] McKay, B. D., and Radziszowski, S. P., $R(4, 5) = 25$. *J. Graph Theory* 19 (1995), 309–322.

[MZ] McKay, B.D., and Zhang, K. M., The Value of the Ramsey Number *R(3,8)*. *J. Graph Theory* 16 (1992), 99–105.

[Ne] Nelsen, R. B., *Proof Without Words II*. Mathematical Association of America, Washington D.C. 2000.

[N] Nilli, A., On Borsuk's Problem. *Jerusalem Combinatorics '93: Papers from the International Conference on Combinatorics held in Jerusalem, May 9–17, 1993* (H. Barcelo and G. Kalai, eds.) Amer. Math. Soc., Providence, RI, 1994, 209–210.

[P] Pál, J. F., Ein Minimumproblem für Ovale. *Math. Ann.* 83 (1921), 311–319 (German).

[Pi] Pikhurko, O., Borsuk's Conjecture Fails in Dimensions 321 and 322, arXiv:math/0202112v1 [math.CO] February 12, 2002.

[Poi] *Materialy Konferenzii "Poisk-97,"* (Materials of the Conference "Poisk-97"). Moscow, 1997, (Russian).

[R] Radon, J., Mengen konvexer Körper, die einen gemeinsamen Punkt enthalten. *Math. Ann.* 83 (1921), 113–115 (German).

[Rad] Radziszowski, S. P., Small Ramsey Numbers, revision #11. *Electronic Journal of Combinatorics*, http://www.cs.rit.edu/~spr/ElJC/abs.pdf, August 1, 2006.

[Ra1] Raigorodski, A. M., On the dimension in Borsuk's problem. *Russian Math. Surveys* 52 (1997), 1324–1325.

[Ra2] Raigorodskii, A. M., The Borsuk partition problem: The seventieth anniversary. *Mathematical Intelligencer* 26(3) (2004), 4–12.

[Rai] Raiskii, D. E., Realizing of all distances in a decomposition of the space R^n into $n + 1$ Parts. *Mat. Zametki* 7 (1970), 319–323 (Russian). English translation: *Math. Notes* 7 (1970), 194–196.

[Ro] Rogers, C. A., Symmetrical sets of constant width and their partitions, *Mathematika* 18 (1971), 105–111.

[S1] Soifer, A., *Mathematics as Problem Solving.* Center for Excellence in Mathematical Education, Colorado Springs, CO, 1987.

[S2] Soifer, A., *How Does One Cut a Triangle?* Center for Excellence in Mathematical Education, Colorado Springs, CO, 1990.

[S3] Soifer, A., Kletchatye doski i polyomino (Checker boards and polyominoes). *Kvant* 11 (1972), 2–10 (Russian).

[S4] Soifer, A., A Six-Coloring of the Plane, *J. Combin, Theory, Ser. A* 61 (2) (1992), 292–294.

[S5] Soifer, A., Six-realizable set X_6, *Geombinatorics* III(4) (1994), 140–145.

[S6] Soifer, A., An infinite class of 6-colorings of the plane, *Congressus Numerantium* 101 (1994), 83–86.

[S7] Soifer, A., *The Mathematical Coloring Book: Mathematics of Coloring and the Colorful Life of Its Creators*, Springer, New York, 2009.

[S8] Soifer, A., *Colorado Mathematical Olympiad: The First Ten Years and Further Explorations,* Center for Excellence in Mathematical Education, Colorado Springs, 1994.

[S9] Soifer, A., *Mathematics as Problem Solving.* 2nd expanded edition, Springer, New York, 2009.

[S10] Soifer, A., *How Does One Cut a Triangle?* 2nd expanded edition, Springer, New York, 2009.

[S11] *Colorado Mathematical Olympiad: The First 20 Years and Further Explorations,* Springer, New York 2011, to appear.

[S12] Soifer, A., *Life and Fate: In Search of Van der Waerden*, Springer, New York, 2011, to appear.

[SL] Soifer, A., and Lozansky, E., Pigeons in every pigeonhole. *Quantum,* January, 1990, 25–26, 32.

[SS] Soifer, A., and Slobodnik, S. G., Problem M236. *Kvant,* 12 (1973), 29 (Russian).

[Spe3] Spencer, T., and Radziszowski, S. P., Personal communication, University of Nebraska at Omaha, 1993.

[Spe4] Spencer, T., *Upper Bounds for Ramsey Numbers*, Manuscript, (1994).

[St1] Starr, N., Tromino Tiling Deficient Cubes of Side Length 2nd. *Geombinatorics* XVIII(3) (2008), 72–87.

[St2] Starr, N., Tromino Tiling Deficient Cubes of Any Side Length, article 0806.0524 in http://arXiv.org/archive/math.

[SN] Szökefalvi-Nagy, B., Ein Satz über Parallel-verschiebungen konvexer Körper. *Ada Sci. Math.* 15 (1954), 169–177 (German).

[T] Tabachnikov, S. L., Considerations of continuity. *Quantum*, May 1990, 8–12.

[W] Wagner, K., Bemerkungen zum Vierfarben problem, *Jber. Deutsch. Math. Verein.* (1936), 21–22 (German).

[Wa] Walkup, D. W., Covering a rectangle with T-tetrominoes. *Amer. Math. Monthly* 72(9) (1965), 986–988.

[We] Weißbach, B., Sets with large Borsuk number. *Beiträge Algebra Geom.* 41 (2000), 417–423.

[Wo] Woodall, D. R., Distances realized by sets covering the plane. *J. Combin. Theory Ser. A* 14 (1973), 187–200.

[XXER] Xu X., Xie Z., Exoo, G., and Radziszowski, S. P., Constructive Lower Bounds on Classical Multicolor Ramsey Numbers. *Electronic J. Combinatorics,* #R35(11) http://www.combinatorics.org/ (2004).

[XXR] Xu, X., Xie, Z, and Radziszowski, S. P., A Constructive Approach for the Lower Bounds on the Ramsey Numbers $R(s,t)$, *J. Graph Theory* 47 (2004), 231–239.

[Y] Yoneyama, K., Theory of continuous set of points. *Tôhoku Math. J.* 12 (1917), 43–158.

[YB] Yaglom, I. M., and Boltjanski, V. G., *Vypuklye Figury* (Convex Figures), GITTL, Moscow, 1951 (Russian). English translation: *Convex Figures*, Holt, Rinehart and Winston, New York, 1961.

Index

Notations

The etudes presented here are not simply those of Czerny, but are better compared to the etudes of Chopin, not only technically demanding and addressed to a variety of specific skills, but at the same time possessing an exceptional beauty that characterizes the best of art ... Keep this book at hand as you plan your next problem solving seminar.
– DON CHAKERIAN in The American Mathematical Monthly

Alexander Soifer, Photograph by Mark S. Soifer

Alexander Soifer's *Geometrical Etudes in Combinatorial Mathematics* is concerned with beautiful mathematics, and it will likely occupy a special and permanent place in the mathematical literature, challenging and inspiring both novice and expert readers with surprising and exquisite problems and theorems... He conveys the joy of discovery as well as anyone, and he has chosen a topic that will stand the test of time.

– CECIL ROUSSEAU

Each time I looked at "Geometric Etudes in Combinatorial Geometry" I found something that was new and surprising to me, even after more than fifty years working in combinatorial geometry.

– BRANKO GRÜNBAUM

All of Alexander Soifer's books can be viewed as excellent and artful entrees to mathematics in the MAPS mode. Different people will have different preferences among them, but here is something that *Ge-*

ometric Etudes does better than the others: after bringing the reader into a topic by posing interesting problems, starting from a completely elementary level, it then goes deep.

– PETER D. JOHNSON JR.

This interesting and delightful book by two well-known geometers is written both for mature mathematicians interested in somewhat unconventional geometric problems and especially for talented young students who are interested in working on unsolved problems which can be easily understood by beginners and whose solutions perhaps will not require a great deal of knowledge but may require a great deal of ingenuity . . . I recommend this book very warmly.

– PAUL ERDŐS

Alexander Soifer and Vladimir Boltyanski have produced a fascinating book, filled with material I have not seen before in any book.

– MARTIN GARDNER

Throughout the text, the authors show great mastery of the topics discussed. Their infectious enthusiasm for opening our eyes to the beauties of the worlds of geometry and combinatorics should make this book attractive to a wide audience. It is to be hoped that the following pages will bring the joy of understanding, seeing, and discovering geometry to many of our young people.

– BRANKO GRÜNBAUM